Applied Data Mining
for Business and Industry

Applied Data Mining for Business and Industry

Second Edition

PAOLO GIUDICI

Department of Economics, University of Pavia, Italy

SILVIA FIGINI

Faculty of Economics, University of Pavia, Italy

WILEY

A John Wiley and Sons, Ltd., Publication

Library of Congress Cataloging-in-Publication Data

Giudici, Paolo.
  Applied data mining for business and industry / Paolo Giudici, Silvia Figini. – 2nd ed.
    p. cm.
  Includes bibliographical references and index.
  ISBN 978-0-470-05886-2 (cloth) – ISBN 978-0-470-05887-9 (pbk.)
  1. Data mining. 2. Business–Data processing. 3. Commercial statistics. I. Figini, Silvia. II. Title.
  QA76.9.D343G75 2009
  005.74068—dc22

                                                  2009008334

A catalogue record for this book is available from the British Library

ISBN: 978-0-470-05886-2 (Hbk)
ISBN: 978-0-470-05887-9 (Pbk)

Typeset in 10/12 Times-Roman by Laserwords Private Limited, Chennai, India
Printed and bound in Great Britain by TJ International, Padstow, Cornwall, UK

Contents

CHAPTER 1

Introduction

From an operational point of view, data mining is an integrated process of data analysis that consists of a series of activities that go from the definition of the objectives to be analysed, to the analysis of the data up to the interpretation and evaluation of the results. The various phases of the process are as follows:

Definition of the objectives for analysis. It is not always easy to define statistically the phenomenon we want to analyse. In fact, while the company objectives that we are aiming for are usually clear, they can be difficult to formalise. A clear statement of the problem and the objectives to be achieved is is of the utmost importance in setting up the analysis correctly. This is certainly one of the most difficult parts of the process since it determines the methods to be employed. Therefore the objectives must be clear and there must be no room for doubt or uncertainty.

Selection, organisation and pre-treatment of the data. Once the objectives of the analysis have been identified it is then necessary to collect or select the data needed for the analysis. First of all, it is necessary to identify the data sources. Usually data is taken from internal sources that are cheaper and more reliable. This data also has the advantage of being the result of the experiences and procedures of the company itself. The ideal data source is the company data warehouse, a 'store room' of historical data that is no longer subject to changes and from which it is easy to extract topic databases (data marts) of interest. If there is no data warehouse then the data marts must be created by overlapping the different sources of company data.

In general, the creation of data marts to be analysed provides the fundamental input for the subsequent data analysis. It leads to a representation of the data, usually in table form, known as a data matrix that is based on the analytical needs and the previously established aims.

Once a data matrix is available it is often necessary to carry out a process of preliminary cleaning of the data. In other words, a quality control exercise is carried out on the data available. This is a formal process used to find or select variables that cannot be used, that is, variables that exist but are not suitable for analysis. It is also an important check on the contents of the variables and

Applied Data Mining for Business and Industry, 2e   P. Giudici, S. Figini
© 2009 John Wiley & Sons, Ltd

the possible presence of missing or incorrect data. If any essential information is missing it will then be necessary to supply further data. (See Agresti (1990).

Exploratory analysis of the data and their transformation. This phase involves a preliminary exploratory analysis of the data, very similar to on-line analytical process (OLAP) techniques. It involves an initial evaluation of the importance of the collected data. This phase might lead to a transformation of the original variables in order to better understand the phenomenon or which statistical methods to use. An exploratory analysis can highlight any anomalous data, data that is different from the rest. This data will not necessarily be eliminated because it might contain information that is important in achieving the objectives of the analysis. We think that an exploratory analysis of the data is essential because it allows the analyst to select the most appropriate statistical methods for the next phase of the analysis. This choice must consider the quality of the available data. The exploratory analysis might also suggest the need for new data extraction, if the collected data is considered insufficient for the aims of the analysis.

Specification of statistical methods. There are various statistical methods that can be used, and thus many algorithms available, so it is important to have a classification of the existing methods. The choice of which method to use in the analysis depends on the problem being studied or on the type of data available. The data mining process is guided by the application. For this reason, the classification of the statistical methods depends on the analysis's aim. Therefore, we group the methods into two main classes corresponding to distinct/different phases of the data analysis.

- **Descriptive methods.** The main objective of this class of methods (also called symmetrical, unsupervised or indirect) is to describe groups of data in a succinct way. This can concern both the observations, which are classified into groups not known beforehand (cluster analysis, Kohonen maps) as well as the variables that are connected among themselves according to links unknown beforehand (association methods, log-linear models, graphical models). In descriptive methods there are no hypotheses of causality among the available variables.

- **Predictive methods.** In this class of methods (also called asymmetrical, supervised or direct) the aim is to describe one or more of the variables in relation to all the others. This is done by looking for rules of classification or prediction based on the data. These rules help predict or classify the future result of one or more response or target variables in relation to what happens to the explanatory or input variables. The main methods of this type are those developed in the field of machine learning such as neural networks (multilayer perceptrons) and decision trees, but also classic statistical models such as linear and logistic regression models.

Analysis of the data based on the chosen methods. Once the statistical methods have been specified they must be translated into appropriate algorithms for computing the results we need from the available data. Given the wide range of specialised and non-specialised software available for data mining, it is not necessary to develop ad hoc calculation algorithms for the most 'standard' applications. However, it is important that those managing the data mining process have a good understanding of the different available methods as well as of the different software solutions, so that they can adapt the process to the specific needs of the company and can correctly interpret the results of the analysis.

Evaluation and comparison of the methods used and choice of the final model for analysis. To produce a final decision it is necessary to choose the best 'model' from the various statistical methods available. The choice of model is based on the comparison of the results obtained. It may be that none of the methods used satisfactorily achieves the analysis aims. In this case it is necessary to specify a more appropriate method for the analysis. When evaluating the performance of a specific method, as well as diagnostic measures of a statistical type, other things must be considered such as the constraints on the business both in terms of time and resources, as well as the quality and the availability of data. In data mining it is not usually a good idea to use just one statistical method to analyse data. Each method has the potential to highlight aspects that may be ignored by other methods.

Interpretation of the chosen model and its use in the decision process. Data mining is not only data analysis, but also the integration of the results into the company decision process. Business knowledge, the extraction of rules and their use in the decision process allow us to move from the analytical phase to the production of a decision engine. Once the model has been chosen and tested with a data set, the classification rule can be generalised. For example, we will be able to distinguish which customers will be more profitable or to calibrate differentiated commercial policies for different target consumer groups, thereby increasing the profits of the company.

Having seen the benefits we can get from data mining, it is crucial to implement the process correctly in order to exploit it to its full potential. The inclusion of the data mining process in the company organisation must be done gradually, setting out realistic aims and looking at the results along the way. The final aim is for data mining to be fully integrated with the other activities that are used to support company decisions. This process of integration can be divided into four phases:

- **Strategic phase.** In this first phase we study the business procedures in order to identify where data mining could be more beneficial. The results at the end of this phase are the definition of the business objectives for a pilot data mining project and the definition of criteria to evaluate the project itself.
- **Training phase.** This phase allows us to evaluate the data mining activity more carefully. A pilot project is set up and the results are assessed using the objectives and the criteria established in the previous phase. A fundamental aspect of the implementation of a data mining procedure is the choice of the pilot project. It must be easy to use but also important enough to create interest.
- **Creation phase.** If the positive evaluation of the pilot project results in implementing a complete data mining system it will then be necessary to

establish a detailed plan to reorganise the business procedure in order to include the data mining activity. More specifically, it will be necessary to reorganise the business database with the possible creation of a data warehouse; to develop the previous data mining prototype until we have an initial operational version and to allocate personnel and time to follow the project.

- **Migration phase.** At this stage all we need to do is to prepare the organisation appropriately so that the data mining process can be successfully integrated. This means teaching likely users the potential of the new system and increasing their trust in the benefits that the system will bring to the company. This means constantly evaluating (and communicating) the results obtained from the data mining process.

PART I

Methodology

CHAPTER 2

Organisation of the data

Data analysis requires the data to be organised into an ordered database. We will not discuss how to create a database in this book. The way in which the data is analysed depends on how the data is organised within the database. In our information society there is an abundance of data which calls for an efficient statistical analysis. However, an efficient analysis assumes and requires a valid organisation of the data.

It is of strategic importance for all medium-sized and large companies to have a unified information system, called a data warehouse, that integrates, for example, the accounting data with the data arising from the production process, the contacts with the suppliers (supply chain management), the sales trends and the contacts with the customers (customer relationship management). This system provides precious information for business management. Another example is the increasing diffusion of electronic trade and commerce and, consequently, the abundance of data about web sites visited together with payment transactions. In this case it is essential for the service supplier to understand who the customers are in order to plan offers. This can be done if the transactions (which correspond to clicks on the web) are transferred to an ordered database that can later be analysed.

Furthermore, since the information which can be extracted from a data mining process (data analysis) depends on how the data is organised it is very important that the data analysts are also involved in setting up the database itself. However, frequently the analyst finds himself with a database that has already been prepared. It is then his/her job to understand how it has been set up and how it can be used to achieve the stated objectives. When faced with poorly set-up databases it is a good idea to ask for these to be reviewed rather than trying to laboriously extract information that might ultimately be of little use.

In the remainder of this chapter we will describe how to transform the database so that it can be analysed. A common structure is the so-called data matrix. We will then consider how sometimes it is a good idea to transform a data matrix in terms of binary variables, frequency distributions, or in other ways. Finally, we will consider examples of more complex data structures.

2.1 Statistical units and statistical variables

From a statistical point of view, a database should be organised according to two principles: the statistical units, the elements in the reference population that

Applied Data Mining for Business and Industry, 2e P. Giudici, S. Figini
© 2009 John Wiley & Sons, Ltd

are considered important for the aims of the analysis (for example, the supply companies, the customers, or the people who visit the site); and the statistical variables, characteristics measured for each statistical unit (for example, if the customer is the statistical unit, customer characteristics might include the amounts spent, methods of payment and socio-demographic profiles).

The statistical units may be the entire reference population (for example, all the customers of the company) or just a sample. There is a large body of work on the statistical theory of sampling and sampling strategies, but we will not go into details here (see Barnett, 1974).

Working with a representative sample rather than the entire population may have several advantages. On the one hand it can be expensive to collect complete information on the entire population, while on the other hand the analysis of large data sets can be time-consuming, in terms of analysing and interpreting the results (think, for example, about the enormous databases of daily telephone calls which are available to mobile phone companies).

The statistical variables are the main source of information for drawing conclusions about the observed units which can then be extended to a wider population. It is important to have a large number of statistical variables; however, such variables should not duplicate information. For example, the presence of the customers' annual income may make the monthly income variable superfluous.

Once the units and the variables have been established, each observation is related to a statistical unit, and, correspondingly, a distinct value (level) for each variable is assigned. This process leads to a data matrix.

Two different types of variables arise in a data matrix: qualitative and quantitative. Qualitative variables are typically expressed verbally, leading to distinct categories. Some examples of qualitative variables include sex, postal codes, and brand preference.

Qualitative variables can be sub-classified into nominal, if their distinct categories appear without any particular order, or ordinal, if the different categories are ordered. Measurement at a nominal level allows us to establish a relation of equality or inequality between the different levels $(=, \neq)$. Examples of nominal measurements are the colour of a person's eyes and the legal status of a company. The use of ordinal measurements allows us to establish an ordered relation between the different categories. More precisely, we can affirm which category is bigger or better $(=, >, <)$ but we cannot say by how much. Examples of ordinal measurements are the computing skills of a person and the credit rate of a company.

Quantitative variables, on the other hand, are numerical – for example age or income. For these it is also possible to establish connections and numerical relations among their levels. They can be classified into discrete quantitative variables, when they have a finite number of levels (for example, the number of telephone calls received in a day), and continuous quantitative variables, if the levels cannot be counted (for example, the annual revenues of a company).

Note that very often the levels of ordinal variables are 'labelled' with numbers. However, this labelling does not make the variables into quantitative ones.

Once the data and the variables have been classified into the four main types (qualitative nominal and ordinal, quantitative discrete and continuous), the database must be transformed into a structure which is ready for a statistical analysis, the data matrix. The data matrix is a table that is usually two-dimensional, where the rows represent the n statistical units considered and the columns represent the p statistical variables considered. Therefore the generic element (i, j) of the matrix $i = 1, \ldots, n$ and $j = 1, \ldots, p$ is a classification of the statistical unit i according to the level of the jth variable.

The data matrix is where data mining starts. In some cases, as in, for example, a joint analysis of quantitative variables, it acts as the input of the analysis phase. In other cases further pre-processing is necessary. This leads to tables derived from data matrices. For example, in the joint analysis of qualitative variables it is a good idea to transform the data matrix into a contingency table. This is a table with as many dimensions as the number of qualitative variables that are in the data set. We shall discuss this point in more detail in the context of the representation of the statistical variables in frequency distributions.

2.2 Data matrices and their transformations

The initial step of a good statistical data analysis has to be exploratory. This is particularly true of applied data mining, which essentially consists of searching for relationships in the data at hand, not known a priori. Exploratory data analysis is usually carried out through computationally intensive graphical representations and statistical summary measures, relevant for the aims of the analysis.

Exploratory data analysis might thus seem, on a number of levels, equivalent to data mining itself. There are two main differences, however. From a statistical point of view, exploratory data analysis essentially uses descriptive statistical techniques, while data mining, as we will see, can use both descriptive and inferential methods, the latter being based on probabilistic methods. Also there is a considerable difference in the purpose of the two analyses. The prevailing purpose of an exploratory analysis is to describe the structure and the relationships present in the data, perhaps for subsequent use in a statistical model. The purpose of a data mining analysis is the production of decision rules based on the structures and models that describe the data. This implies, for example, a considerable difference in the use of alternative techniques. An exploratory analysis often consists of several different exploratory techniques, each one capturing different potentially noteworthy aspects of the data. In data mining, on the other hand, the various techniques are evaluated and compared in order to choose one for later implementation as a decision rule. A further discussion of the differences between exploratory data analysis and data mining can be found in Coppi (2002).

The next chapter will explain exploratory data analysis. First, we will discuss univariate exploratory analysis, the examination of available variables one at a time. Even though the observed data is multidimensional and, therefore, we need to consider the relationships between the available variables, we can gain a great deal of insight by examining each variable on its own. We will then consider multivariate aspects, starting with bivariate relationships.

Often it seems natural to summarise statistical variables with a frequency distribution. As it happens for all procedures of this kind, the summary makes the analysis and presentation of the results easier but it also naturally leads to a loss of information. In the case of qualitative variables the summary is justified by the need to be able to carry out quantitative analysis on the data. In other situations, such as in the case of quantitative variables, the summary is done essentially with the aim of simplifying the analysis.

We start with the analysis of a single variable (univariate analysis). It is easier to extract information from a database by starting with univariate analysis and then going on to a more complicated analysis of multivariate type. The determination of the univariate distribution frequency starting off from the data matrix is often the first step of a univariate exploratory analysis. To create a frequency distribution for a variable it is necessary to establish the number of times each level appears in the data. This number is called the absolute frequency. The levels and their frequency together give the frequency distribution.

Multivariate frequency distributions are represented in contingency tables. To make our explanation clearer we will consider a contingency table with two dimensions. Given such a data structure it is easy to calculate descriptive measures of association (odds ratios) or dependency (chi-square).

The transformation of the data matrix into univariate and multivariate frequency distributions is not the only possible transformation. Other transformations can also be very important in simplifying the statistical analysis and/or the interpretation of the results. For example, when the p variables of the data matrix are expressed in different units of measure it is a good idea to standardise the variables, subtracting the mean of each one and dividing it by the square root of its variance. The variable thus obtained has mean equal to zero and variance equal to unity.

The transformation of data is also a way of solving quality problems because some data may be missing or may have anomalous values (outliers). Two main approaches are used with missing data: (a) it may be removed; (b) it may be substituted it by means of an appropriate function of the remaining data. A further problem occurs with outliers. Their identification is often itself a reason for data mining. Unlike what happens with missing data, the discovery of an anomalous value requires a formal statistical analysis, and usually it cannot be eliminated. For example, in the analysis of fraud detection (related to telephone calls or credit cards, for example), the aim of the analysis is to identify suspicious behaviour. For more information about the problems related to data quality, see Han and Kamber (2001).

2.3   Complex data structures

The application aims of data mining may require a database not expressible in terms of the data matrix we have used up to now. For example, there are often other aspects of data collection to consider, such as time and/or space. In this kind of application the data is often presented aggregated or divided (for example, into periods or regions) and this is an important aspect that must be considered (on this topic see Diggle *et al.*, 1994).

The most important case refers to longitudinal data – for example, the comparison in n companies of the p budget variables in q subsequent years. In this case there will be a three-way matrix that can be described by three dimensions: n statistical units, p statistical variables and q time periods. Another important example of data matrices with more than two dimensions concerns the presence of data related to different geographic areas. In this case, as in the previous one, there is a three-way matrix with space as the third dimension – for example, the sales of a company in different regions or the satellite surveys of the environmental characteristics of different regions. In such cases, data mining should use times series methods (for an introduction see Chatfield, 1996) or spatial statistics (for an introduction see Cressie, 1991).

However, more complex data structures may arise. Three important examples are text data, web data, and multimedia data. In the first case the available database consists of a library of text documents, usually related to each other. In the second case, the data is contained in log files that describe what each visitor does at a web site during a session. In the third case, the data can be made up of texts, images, sounds and other forms of audio-visual information that is typically downloaded from the internet and that describes an interaction with the web site more complex than the previous example. Obviously this type of data implies a more complex analysis. The first challenge in analysing this kind of data is how to organize it. This has become an important research topic in recent years (see Han and Kamber, 2001). In Chapter 6 we will show how to analyze web data contained in a log file.

Another important type of complex data structure arises from the integration of different databases. In modern applications of data mining it is often necessary to combine data that come from different sources, for example internal and external data about operational losses, as well as perceived expert opinions (as in Chapter 12). For further discussion about this problem, also known as data fusion, see Han and Kamber (2001).

Finally, let us mention that some data are now observable in continuous rather than discrete time. In this case the observations for each variable on each unit are a function rather than a point value. Important examples include monitoring the presence of polluting atmospheric agents over time and surveys on the quotation of various financial shares. These are examples of continuous time stochastic processes which are described, for instance, in Hoel *et al.* (1972).

2.4   Summary

In this chapter we have given an introduction to the organisation and structure of the databases that are the object of the data mining analysis. The most important point is that the planning and creation of the database cannot be ignored but it is one of the most important data mining phases. We see data mining as a process consisting of design, collection and data analysis. The main objectives of the data mining process are to provide companies with useful/new knowledge in the sphere of business intelligence. The elements that are part of the creation of the database or databases and the subsequent analysis are closely interconnected. Although the chapter summarises the important aspects given the statistical rather than computing nature of the book, we have tried to provide an introductory overview.

We conclude this chapter with some useful references for the topics introduced in this chapter. The chapter started with a description of the various ways in which we can structure databases. For more details on these topics, see Han and Kamber (2001), from a computational point of view; and Berry and Linoff (1997, 2000) from a business-oriented point of view. We also discussed fundamental classical topics, such as measurement scales. This leads to an important taxonomy of the statistical variables that is the basis of the operational distinction of data mining methods that we adopt here. Then we introduced the concept of data matrices. The data matrix allows the definition of the objectives of the subsequent analysis according to the formal language of statistics. For an introduction to these concepts, see Hand *et al.* (2001). We also introduced some transformations on the data matrix, such as the calculation of frequency distributions, variable transformations and the treatment of anomalous or missing data. For all these topics, which belong the preliminary phase of data mining, we refer the reader to Hand *et al.* (2001), from a statistical point of view, and Han and Kamber (2001), from a computational point of view. Finally, we briefly described complex data structures; for more details the reader can also consult Hand *et al.* (2001) and Han and Kamber (2001).

CHAPTER 3

Summary statistics

In this chapter we introduce univariate summary statistics used to summarize the distribution of univariate variables. We then consider multivariate distributions, starting with summary statistics for bivariate distributions and then moving on to multivariate exploratory analysis of qualitative data. In particular, we compare some of the numerous summary measures available in the statistical literature. Finally, in consideration of the difficulty in representing and displaying high-dimensional data and results, we discuss a popular statistical method for reducing dimensionality, principal components analysis.

3.1 Univariate exploratory analysis

3.1.1 Measures of location

The most common measure of location is the (arithmetic) mean, which can be computed only for quantitative variables. The mean of a set $x_1, x_2, \ldots, x_N$ of N observations is given by

$$\bar{x} = \frac{x_1 + x_2 + \cdots + x_N}{N} = \sum \frac{x_i}{N}.$$

We note that, in the calculation of the arithmetic mean, the biggest observations, can counterbalance and even overpower the smallest ones. Since, all the observations are used in the computation of the mean, its value can be influenced by outlying observations In financial data where extreme observations are common, this happens often and, therefore, alternatives to the mean are probably preferable as measures of location.

The previous expression for the arithmetic mean can be calculated on the data matrix. Table 3.1 shows the structure of a data matrix and Table 3.2 an example. When univariate variables are summarised with the frequency distribution, the arithmetic mean can also be calculated directly from the frequency distribution. This computation leads, of course, to the same mean value and saves computing time. The formula for computing the arithmetic mean from the frequency distribution is given by

$$\bar{x} = \sum x_i^* p_i.$$

Applied Data Mining for Business and Industry, 2e   P. Giudici, S. Figini
© 2009 John Wiley & Sons, Ltd

Table 3.1   Data matrix.

	1	j	p
1	$X_{1,1}$	$X_{1,j}$	$X_{1,p}$
⋮			
i	$X_{i,1}$	$X_{i,j}$	$X_{i,p}$
⋮			
n	$X_{n,1}$	$X_{n,j}$	$X_{n,p}$

Table 3.2   Example of a data matrix.

	Y	X_1	X_2	...	X_5	...	...	...	...	...	X_{20}
N 1	1	1	18	...	1049		...				1
...											
N 34	1	4	24	...	1376		...				1
...											
...											
N 1000	0	1	30	...	6350		...				1

This formula is known as the weighted arithmetic mean, where the x_i^* indicate the distinct levels that the variable can take on and p_i is the relative frequency of each of these levels.

We list below the most important properties of the arithmetic mean:

- The sum of the deviations from the mean is zero: $\sum(x_i - \overline{x}) = 0$.
- The arithmetic mean is the constant that minimises the sum of the squares of the deviations of each observation from the constant itself: $\min_a \sum(x_i - a)^2 = \overline{x}$.
- The arithmetic mean is a linear operator: $N^{-1}\sum(a + bx_i) = a + b\overline{x}$.

A second simple measure of position or location is the modal value or mode. The mode is a measure of location computable for all kinds of variables, including qualitative nominal ones. For qualitative or discrete quantitative characters, the mode is the level associated with the greatest frequency. To estimate the mode of a continuous variable, we generally discretize the values that the variables assumes in intervals and compute the mode as the interval with the maximum density (corresponding to the maximum height of the histogram). To obtain a unique mode the convention is to use the middle value of the mode's interval.

Finally, another important measure of position is the median. Given an ordered sequence of observations, the median is the value such that half of the observations are greater than and half are smaller than it. The median can be computed for quantitative variables and ordinal qualitative variables. Given N observations in non-decreasing order the median is:

- if N is odd, the observation which occupies position $(N+1)/2$;
- if N is even, the mean of the observations that occupy positions $N/2$ and $(N/2)+1$.

Note that the median remains unchanged if the smallest and largest observations are substituted with any other value that is lower (or greater) than the median. For this reason, unlike the mean, anomalous or extreme values do not influence the median value.

The comparison between the mean and the median can be usefully employed to detect the asymmetry of a distribution. Figure 3.1 shows three different frequency distributions, which are skewed to the right, symmetric, and skewed to the left, respectively.

As a generalisation of the median, one can consider the values that break the frequency distribution into parts, of preset frequencies or percentages. Such values are called quantiles or percentiles. Of particular interest are the quartiles, which correspond to the values which divide the distribution into four equal parts. The first, second, and third quartiles, denoted by q_1, q_2, q_3, are such that the overall relative frequency with which we observe values less than q_1 is 0.25, less than q_2 is 0.5 and less than q_3 is 0.75. Observe that q_2 coincides with the median.

3.1.2 Measures of variability

In addition to the measures giving information about the position of a distribution, it is important also to summarise the dispersion or variability of the distribution of a variable. A simple indicator of variability is the difference between the maximum value and the minimum value observed for a certain variable, known as the range. Another measure that can be easily computed is interquartile range (IQR), given by the difference between the third and first quartiles, $q_3 - q_1$. While the range is highly sensitive to extreme observations, the IQR is a robust measure of spread for the same reason that the median is a robust measure of location.

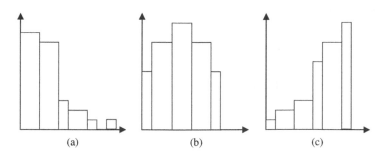

(a)   (b)   (c)

Figure 3.1 Frequency distributions (histograms) describing symmetric and asymmetric distributions: (a) mean > median; (b) mean = median; (c) mean < median.

However, such indexes are not often used. The most commonly used measure of variability for quantitative data is the variance. Given a set $x_1, x_2, \ldots, x_N$ of N observations of a quantitative variable X, with arithmetic mean $\bar{x}$, the variance is defined by

$$\sigma^2(X) = \frac{1}{N} \sum (x_i - \bar{x})^2,$$

which is approximately the average squared deviation from the mean. When calculated on a sample rather than the whole population it is also denoted by s^2. Note that when all the observations assume the same value the variance is zero. Unlike the mean, the variance is not a linear operator, since $\mathrm{Var}(a + bX) = b^2 \mathrm{Var}(X)$.

The units of measure of the variance are the the units of measure of X squared. That is, if X is measured in metres, then the variance is measured in metres squared. For this reason the square root of the variance, known as the standard deviation, is preferred. Furthermore, to facilitate comparisons between different distributions, the coefficient of variation (CV) is often used. The CV equals the standard deviation divided by the absolute value of the arithmetic mean of the distribution (obviously defined only when the latter is non-zero). The CV is a unitless measure of spread.

3.1.3   Measures of heterogeneity

The measures of variability discussed in the previous section cannot be computed for qualitative variables. It is therefore necessary to develop an index able to measure the dispersion of the distribution also for this type of data. This is possible by resorting to the concept of heterogeneity of the observed distribution of a variable. Tables 3.3 and 3.4 show the structure of a frequency distribution, in terms of absolute and relative frequencies, respectively.

Consider the general representation of the frequency distribution of a qualitative variable with k levels (Table 3.4). In practice it is possible to have two extreme situations between which the observed distribution will lie. Such situations are the following:

- Null heterogeneity, when all the observations have X equal to the same level. That is, if $p_i = 1$ for a certain i, and $p_i = 0$ for the other $k - 1$ levels.

Table 3.3 Univariate frequency distribution.

Levels	Absolute frequencies
x_1^*	n_1
x_2^*	n_2
$\vdots$	$\vdots$
x_k^*	n_k

Table 3.4 Univariate relative frequency distribution.

Levels	Relative frequencies
x_1^*	p_1
x_2^*	p_2
$\vdots$	$\vdots$
x_k^*	p_k

- Maximum heterogeneity, when the observations are uniformly distributed amongst the k levels, that is $p_i = 1/k$ for all $i = 1, \ldots, k$.

A heterogeneity index will have to attain its minimum in the first situation and its maximum in the second. We now introduce two indexes that satisfy such conditions.

The Gini index of heterogeneity is defined by

$$G = 1 - \sum_{i=1}^{k} p_i^2.$$

It can be easily verified that the Gini index is equal to 0 in the case of perfect homogeneity and $1 - 1/k$ in the case of maximum heterogeneity. To obtain a 'normalised' index, which assumes values in the interval [0,1], the Gini index can be rescaled by its maximum value, giving the following relative index of heterogeneity:

$$G' = \frac{G}{(k-1)/k}.$$

The second index of heterogeneity is the entropy, defined by

$$E = -\sum_{i=1}^{k} p_i \log p_i.$$

This index equals 0 in the case of perfect homogeneity and $\log k$ in the case of maximum heterogeneity. To obtain a 'normalised' index, which assumes values in the interval [0,1], E can be rescaled by its maximum value, giving the following relative index of heterogeneity:

$$E' = \frac{E}{\log(k)}.$$

3.1.4 Measures of concentration

A statistical concept which is very much related to heterogeneity is that of concentration. In fact, a frequency distribution is said to be maximally concentrated when it has null heterogeneity and minimally concentrated when it has a maximal heterogeneity. It is interesting to examine intermediate situations, where the two concepts find a different interpretation. In particular, the concept of concentration applies to variables measuring transferable goods (both quantitative and ordinal qualitative). The classical example is the distribution of a fixed amount of income among N individuals, which we shall use as a running example.

Consider N non-negative quantities measuring a transferable characteristic placed in non-decreasing order: $0 \leq x_1 \leq \ldots \leq x_N$. The aim is to understand the concentration of the characteristic among the N quantities, corresponding to different observations. Let $N \bar{x} = \sum x_i$ be the total available amount, where $\bar{x}$ is the arithmetic mean. Two extreme situations can arise:

- $x_1 = x_2 = \ldots = x_N = \bar{x}$, corresponding to minimum concentration (equal income across the N units for the running example);
- $x_1 = x_2 = \ldots = x_{N-1} = 0, x_N = N\bar{x}$, corresponding to maximum concentration (only one unit has all the income).

In general, it is of interest to evaluate the degree of concentration, which usually will be between these two extremes. To achieve this aim we will construct a measure of the concentration. Define

$$F_i = \frac{i}{N}, \qquad \text{for } i = 1, \ldots, N,$$

$$Q_i = \frac{x_1 + x_2 + \cdots + x_i}{N\bar{x}} = \frac{\sum_{j=1}^{i} x_j}{N\bar{x}}, \qquad \text{for } i = 1, \ldots, N.$$

For each i, F_i is the cumulative percentage of units considered, up to the ith, while Q_i describes the cumulative percentage of the characteristic that belongs to the same first i units. It can be shown that:

$$0 \leq F_i \leq 1 \; ; \; 0 \leq Q_i \leq 1,$$

$$Q_i \leq F_i,$$

$$F_N = Q_N = 1.$$

Let $F_0 = Q_0 = 0$ and consider the $N+1$ pairs of coordinates $(0,0)$, $(F_1, Q_1), \ldots, (F_{N-1}, Q_{N-1})$, $(1,1)$. If we plot these points in the plane and join them with line segments, we obtain a piecewise linear curve called the concentration curve (Figure 3.2). From the curve one can clearly see the departure of the observed situation from the case of minimal concentration, and, similarly, from the case of maximum concentration, described by a curve almost coinciding with the x-axis (at least until the $(N-1)$th point).

A summary index of concentration is the Gini concentration index, based on the differences $F_i - Q_i$. There are three points to note:

- For minimum concentration, $F_i - Q_i = 0, i = 1, 2, \ldots, N$.

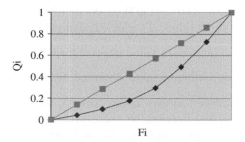

Figure 3.2 Representation of the concentration curve.

- For maximum concentration, $F_i - Q_i = F_i, i = 1, 2, \ldots, N - 1$ and $F_N - Q_N = 0$.
- In general, $0 < F_i - Q_i < F_i, i = 1, 2, \ldots, N - 1$, with the differences increasing as maximum concentration is approached.

The concentration index, denoted by R, is defined by the ratio between the quantity $\sum_{i=1}^{N-1} (F_i - Q_i)$ and its maximum value, equal to $\sum_{i=1}^{N-1} F_i$. Thus,

$$R = \frac{\sum_{i=1}^{N-1} (F_i - Q_i)}{\sum_{i=1}^{N-1} F_i}$$

and R assumes value 0 for minimal concentration and 1 for maximum concentration.

3.1.5 Measures of asymmetry

In order to obtain an indication of the asymmetry of a distribution it may be sufficient to compare the mean and median. If these measures are almost the same, the variable under consideration should have a symmetric distribution. If the mean exceeds the median the data can be described as skewed to the right, while if the median exceeds the mean the data can be described as skewed to the left. Graphs of the data using bar diagrams or histograms are useful for investigating the shape of the variables distribution.

A further graphical tool that permits investigation of the form of a distribution is the boxplot. The box plot, as shown in Figure 3.3, shows the median (Me) and

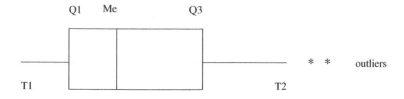

Figure 3.3 A boxplot.

the first and third quartiles (Q1 and Q3) of the distribution of a variable. It also shows the lower and upper limits, T1 and T2, defined by

$$T1 = \max(\text{minimum value observed, } Q1 - 1.5 \text{ IQR}),$$

$$T2 = \min(\text{minimum value observed, } Q3 + 1.5 \text{ IQR}).$$

Examination of the boxplot allows us to identify the asymmetry of the distribution of interest. If the distribution were symmetric the median would be equidistant from Q1 and Q3. Otherwise, the distribution would be skewed. For example, when the distance between Q3 and the median is greater than the distance between Q1 and the median, the distribution is skewed to the right. The boxplot also indicates the presence of anomalous observations or outliers. Observations smaller than T1 or greater than T2 can indeed be seen as outliers, at least on an exploratory basis.

We now introduce a summary statistical index than can measures the degree of symmetry or asymmetry of a distribution. The proposed asymmetry index is function of a quantity known as the third central moment of the distribution:

$$\mu_3 = \frac{\sum (x_i - \bar{x})^3}{N}.$$

The index of asymmetry, known as skewness, is then defined by

$$\gamma = \frac{\mu_3}{s^3},$$

where s is the standard deviation. We note that, as it is evident from its definition, the skewness can be obtained only for quantitative variables. In addition, we note that the proposed index can assume any real value (that is, it is not normalised). We observe that if the distribution is symmetric, $\gamma = 0$; if it is skewed to the left, $\gamma < 0$; finally, if it is skewed to the right, $\gamma > 0$.

3.1.6 Measures of kurtosis

When the variables unders study are continuous, it is possible to approximate, or better, to interpolate the frequency distribution (histogram) with a density function. In particular, in the case in which the number of classes of the histogram is very large and the width of each class is limited, it can be assumed that the histogram can be approximated with a normal or Gaussian density function, having a bell shape (see Figure 3.4).

The normal distribution is an important theoretical model frequently used in inferential statistical analysis. Therefore it may be reasonable to construct a statistical index that measures the 'distance' of the observed distribution from the theoretical situation corresponding to perfect normality. A simple index that allows us to check if the examined data follows a normal distribution is the index of kurtosis, defined by

$$\beta = \frac{\mu_4}{\mu_2^2}$$

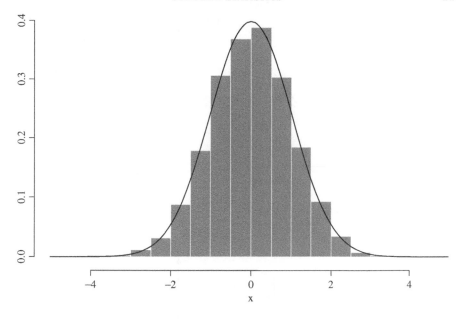

Figure 3.4 Normal approximation to the histogram.

where

$$\mu_4 = \frac{\sum (x_i - \overline{x})^4}{N} \quad \text{and} \quad \mu_2 = \frac{\sum (x_i - \overline{x})^2}{N}.$$

Note that the proposed index can be obtained only for quantitative variables and can assume any real positive value. In particular cases, if the variable is perfectly normal, $\beta = 3$. Otherwise, if $\beta < 3$ the distribution is called hyponormal (thinner with respect to the normal distribution having the same variance, so there is a lower frequency of values very distant from the mean); and if $\beta > 3$ the distribution is called hypernormal (fatter with respect to the normal distribution, so there is a greater frequency for values very distant from the mean).

There are other graphical tools useful for checking whether the data at hand can be approximated with a normal distribution. The most common is the so-called quantile–quantile (QQ) plot. This is a graph in which the observed quantiles from the observed data are compared with the theoretical ones that would be obtained if the data came exactly from a normal distribution (Figure 3.5). If the points plotted fall near the 45° line passing through the origin, then the observed data have a distribution 'similar' a normal distribution.

To conclude this section on univariate analysis, we note that with most of the popular statistical software packages it is easy to produce the measures and graphs described in this section, together with others.

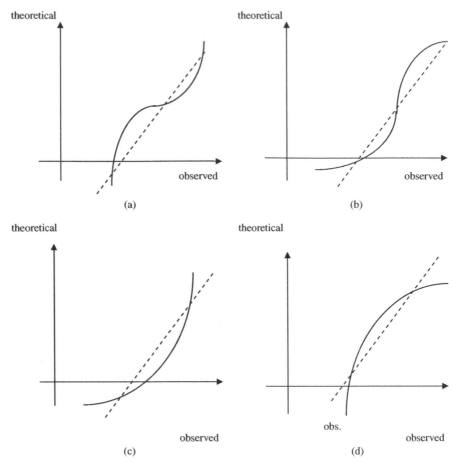

Figure 3.5 Theoretical examples of QQ plots: (a) hyponormal distribution; (b) hypernormal distribution; (c) left asymmetric distribution; (d) right asymmetric distribution.

3.2 Bivariate exploratory analysis of quantitative data

The relationship between two variables can be graphically represented by a scatterplot like that in Figure 3.6. A real data set usually contains more than two variables. In such a case, it is still possible to extract interesting information from the analysis of every possible bivariate scatterplot between all pairs of the variables. We can create a scatterplot matrix in which every element is a scatterplot of the two corresponding variables indicated by the row and the column.

In the same way as for univariate exploratory analysis, it is useful to develop statistical indexes that further summarise the frequency distribution, improving the interpretation of data, even though we may lose some information about the distribution. In the bivariate and, more generally, multivariate case, such indexes allow us not only to summarise the distribution of each data variable, but also

Scatterplot diagram

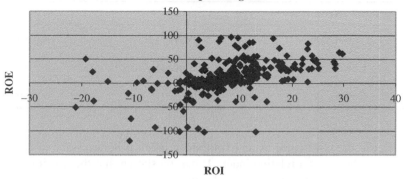

ROI

Figure 3.6 Example of a scatterplot diagram.

to learn about the relationship among variables (corresponding to the columns of the data matrix). In the rest of this section we focus on quantitative variables for which summary indexes are more easily computed. Later, we will see how to develop summary indexes that describe the relationship between qualitative variables.

Concordance is the tendency to observe high (low) values of a variable together with high (low) values of another. Discordance, on the other hand, is the tendency of observing low (high) values of a variable together with high (low) values of the other. The most common summary measure of concordance is the covariance, defined as

$$\text{Cov}(X, Y) = \frac{1}{N} \sum_{i=1}^{N} [x_i - \mu(X)][y_i - \mu(Y)],$$

where $\mu(X)$ and $\mu(Y)$ indicate the mean of the variables X and Y, respectively. The covariance takes positive values if the variables are concordant and negative values if they are discordant. With reference to the scatterplot representation, setting the point $(\mu(X), \mu(Y))$ as the origin, $\text{Cov}(X, Y)$ tends to be positive when most of the observations are in the upper right-hand and lower left-hand quadrants, and negative when most of the observations are in the lower right-hand and upper left-hand quadrants.

The covariance can be directly calculated from the data matrix. In fact, since there is a covariance for each pair of variables, this calculation gives rise to a new data matrix, called the variance–covariance matrix (see Table 3.5). In this matrix the rows and columns correspond to the available variables. The main diagonal contains the variances, while the cells off the main diagonal contain the covariances between each pair of variables. Note that since $\text{Cov}(X_j, X_i) = \text{Cov}(X_i, X_j)$, the resulting matrix will be symmetric.

We remark that the covariance is an absolute index. That is, with the covariance it is possible to identify the presence of a relationship between two quantities but little can be said about the degree of such relationship. In other words, in

Table 3.5   Variance–covariance matrix.

	X_1	...	X_j	...	X_h
X_1	$\text{Var}(X_1)$	...	$\text{Cov}(X_1, X_j)$	...	$\text{Cov}(X_1, X_h)$
...	...	...	...	...	...
X_j	$\text{Cov}(X_j, X_1)$		$\text{Var}(X_j)$	...	...
...	...	...	...	...	...
X_h	$\text{Cov}(X_h, X_1)$	...	...	...	$\text{Var}(X_h)$

order to use the covariance as an exploratory index it is necessary to normalise it, so that it becomes a relative index. It can be shown that the maximum value that $\text{Cov}(X, Y)$ can assume is $\sigma_x \sigma_y$, the product of the two standard deviations of the variables. On the other hand, the minimum value that $\text{Cov}(X, Y)$ can assume is $-\sigma_x \sigma_y$. Furthermore, $\text{Cov}(X, Y)$ takes its maximum value when the observed data lie on a line with positive slope and its minimum value when all the observed data lie on a line with negative slope. In light of this, we define the (linear) correlation coefficient between two variables X and Y as

$$r(X, Y) = \frac{\text{Cov}(X, Y)}{\sigma(X)\sigma(Y)}.$$

The correlation coefficient $r(X, Y)$ has the following properties:

- $r(X, Y)$ takes the value 1 when all the points corresponding to the paired observations lie on a line with positive slope, and it takes the value -1 when all the points lie on a line with negative slope. Due to this property r is known as the *linear* correlation coefficient.
- When $r(X, Y) = 0$ the two variables are not linearly related, that is, X and Y are uncorrelated.
- In general, $-1 \leq r(X, Y) \leq 1$.

As for the covariance, it is possible to calculate all pairwise correlations directly from the data matrix, thus obtaining a correlation matrix (see Table 3.6).

From an exploratory point of view, it is useful to have a threshold-based rule that tells us when the correlation between two variables is 'significantly' different from zero. It can be shown that, assuming that the observed sample

Table 3.6   Correlation matrix.

	X_1	...	X_j	...	X_h
X_1	1	...	$\text{Cor}(X_1, X_j)$	...	$\text{Cor}(X_1, X_h)$
...	...	...	...	...	...
X_j	$\text{Cor}(X_j, X_1)$		1	...	...
...	...	...	...	...	...
X_h	$\text{Cor}(X_h, X_1)$	...	...	...	1

comes from a bivariate normal distribution, the correlation between two variables is significantly different from zero when

$$\left| \frac{r(X, Y)}{\sqrt{1 - r^2(X, Y)}} \sqrt{n - 2} \right| > t_{\alpha/2},$$

where $t_{\alpha/2}$ is the $100(1 - \alpha/2)\%$ percentile of a Student's t distribution with $n - 2$ degrees of freedom, n being the number of observations. For example, for a large sample, and a significance level of $\alpha = 0.05$ (which sets the probability of incorrectly rejecting a null correlation), the threshold is $t_{0.025} = 1.96$.

3.3   Multivariate exploratory analysis of quantitative data

We now show how the use of matrix notation allows us to summarise multivariate relationships among the variables in a more compact way. This also facilitates explanation of multivariate exploratory analysis in general terms, without necessarily going through the bivariate case. In this section we assume that the data matrix contains exclusively quantitative variables. In the next section we will deal with qualitative variables.

Let $\mathbf{X}$ be a data matrix with n rows and p columns. The main summary measures can be expressed directly in terms of matrix operations on $\mathbf{X}$. For example, the arithmetic mean of the variables, described by a p-dimensional vector $\overline{\mathbf{X}}$, can be obtained directly from the data matrix as

$$\overline{\mathbf{X}} = \frac{1}{n} \mathbf{1}\, \mathbf{X},$$

where $\mathbf{1}$ indicates a (row) vector of length n with all elements equal to 1. As previously mentioned, it is often better to standardise the variables in $\mathbf{X}$. To achieve this aim, we first need to subtract the mean from each variable. The matrix containing the deviations from each variable's mean is given by

$$\tilde{\mathbf{X}} = \mathbf{X} - \frac{1}{n} \mathbf{J}\, \mathbf{X},$$

where $\mathbf{J}$ is a $n \times n$ matrix with all the elements equal to 1.

Consider now the variance–covariance matrix, $\mathbf{S}$. This is a $p \times p$ square matrix containing the variance of each variable on the main diagonal. The off-diagonal elements contain the $p(p - 1)/2$ covariances between all the pairs of the p variables. In matrix notation we can write:

$$\mathbf{S} = \frac{1}{n} \tilde{X}' \tilde{X}$$

where $\tilde{X}'$ represents the transpose of $\tilde{\mathbf{X}}$. The (i, j)th element of the matrix is equal to

$$\mathbf{S}_{i,j} = \frac{1}{n} \sum_{\ell=1}^{n} (x_{\ell i} - \overline{x}_i)(x_{\ell j} - \overline{x}_j).$$

$\mathbf{S}$ is symmetric and positive definite, meaning that for any non-zero vector $\mathbf{x}$, $\mathbf{x}'\mathbf{S}\mathbf{x} > 0$.

It may be appropriate, for example in comparing different databases, to summarise the whole variance–covariance matrix with a real number that expresses the 'overall variability' of the system. There are two measures available for this purpose. The first measure, the trace, denoted by tr, is the sum of the elements on the main diagonal of $\mathbf{S}$, the variances of the variables:

$$\text{tr}(\mathbf{S}) = \sum_{s=1}^{p} \sigma_s^2.$$

It can be shown that the trace of $\mathbf{S}$ is equal to the sum of the eigenvalues of $\mathbf{S}$:

$$\text{tr}(\mathbf{S}) = \sum_{s=1}^{p} \lambda_s.$$

A second measure of overall variability is defined by the determinant of $\mathbf{S}$, and it is often called the Wilks generalised variance: $W = |S|$.

In the previous section we saw that it is easy to transform the variance–covariance matrix into the correlation matrix, making the relationships more easily interpretable. The correlation matrix, $\mathbf{R}$, is given by

$$\mathbf{R} = \frac{1}{n}\mathbf{Z}'\mathbf{Z},$$

where $\mathbf{Z} = \tilde{\mathbf{X}}\mathbf{F}$ is a matrix containing the standardised variables and $\mathbf{F}$ is a $p \times p$ matrix that has diagonal elements equal to the reciprocal of the standard deviations of the variables,

$$\mathbf{F} = [\text{diag}(s_{11}, \ldots, s_{pp})]^{-1}.$$

We note that, although the correlation matrix is very informative on the presence of statistical (linear) relationships between the variables of interest, in reality it calculates such relationship marginally for every pair of variables, without taking into account the influence of the other variables on such relationship.

In order to 'filter' the correlations from spurious effects induced by other variables, a useful concept is that of partial correlation. The partial correlation measures the linear relationship between two variables with the others held fixed. Let $r_{ij|REST}$ be the partial correlation observed between the variables X_i and X_j, given *all* the remaining variables, and let $\mathbf{K} = \mathbf{R}^{-1}$, the inverse of the correlation matrix; then the partial correlation is given by

$$r_{ij|\text{REST}} = \frac{-k_{ij}}{[k_{ii}k_{jj}]^{1/2}},$$

where k_{ii}, k_{jj}, and k_{ij} are respectively the (i, i)th, (j, j)th and (i, j)th elements of the matrix $\mathbf{K}$. The importance of reasoning in terms of partial correlations is particularly evident in databases characterised by strong correlations between the variables.

3.4 Multivariate exploratory analysis of qualitative data

We now discuss the exploratory analysis of multivariate data of qualitative type. Hitherto we have used the concept of covariance and correlation as the main measures of statistical relationships among quantitative variables. In the case of ordinal qualitative variables, it is possible to extend the notion of covariance and correlation via the concept of ranks. The correlation between the ranks of two variables is known as the Spearman correlation coefficient. More generally, transforming the levels of the ordinal qualitative variables into the corresponding ranks allows most of the analysis applicable to quantitative data to be extended to the ordinal qualitative case.

However, if the data matrix contains qualitative data at the nominal level the notion of covariance and correlation cannot be used. In this section we consider summary measures for the intensity of the relationships between qualitative variables of any kind. Such measures are known as association indexes. Although suited for qualitative variables, these indexes can be applied to discrete quantitative variables as well (although this entails a loss of explanatory power).

In the examination of qualitative variables a fundamental part is played by the frequencies with which the levels of the variables occur. The usual starting point for the analysis of qualitative variables is the creation or computation of contingency tables (see Table 3.7). We note that qualitative data are often available in the form of a contingency table and not in the data matrix format. To emphasise this difference, we now introduce a slightly different notation.

Given a qualitative variable X which assumes the levels $X_1, \ldots, X_I$, collected in a population (or sample) of n units, the absolute frequency n_i of the level X_i ($i = 1, \ldots, I$) is the number of times that the level X_i is observed in the sample or population. Denote by n_{ij} the frequency associated with the pair of levels (X_i, Y_j), for $i = 1, 2, \ldots, I$ and $j = 1, 2, \ldots, J$, of the variables X and Y. The n_{ij} are also called cell frequencies. Then $n_{i+} = \sum_{j=1}^{J} n_{ij}$ is the marginal frequency of the ith row of the table and represents the total number of observations that assume the ith level of X ($i = 1, 2, \ldots, I$); and $n_{+j} = \sum_{i=1}^{I} n_{ij}$ is the marginal frequency of the jth column of the table and represents the total

Table 3.7 A two-way contingency table.

$X\backslash Y$	y_1^*	y_2^*	$\cdots$	y_j^*	$\cdots$	y_k^*	
x_1^*	$n_{xy}(x_1^*, y_1^*)$	$n_{xy}(x_1^*, y_2^*)$	$\cdots$	$n_{xy}(x_1^*, y_j^*)$	$\cdots$	$n_{xy}(x_1^*, y_k^*)$	$n_x(x_1^*)$
x_2^*	$n_{xy}(x_2^*, y_1^*)$	$n_{xy}(x_2^*, y_2^*)$	$\cdots$	$n_{xy}(x_2^*, y_j^*)$	$\cdots$	$n_{xy}(x_2^*, y_k^*)$	$n_x(x_2^*)$
$\vdots$	$\vdots$	$\vdots$	$\vdots$	$\vdots$	$\vdots$	$\vdots$	$\vdots$
x_i^*	$n_{xy}(x_i^*, y_1^*)$	$n_{xy}(x_i^*, y_2^*)$	$\cdots$	$n_{xy}(x_i^*, y_j^*)$	$\cdots$	$n_{xy}(x_i^*, y_k^*)$	$n_x(x_i^*)$
$\vdots$	$\vdots$	$\vdots$	$\vdots$	$\vdots$	$\vdots$	$\vdots$	$\vdots$
x_h^*	$n_{xy}(x_h^*, y_1^*)$	$n_{xy}(x_h^*, y_2^*)$	$\cdots$	$n_{xy}(x_h^*, y_j^*)$	$\cdots$	$n_{xy}(x_h^*, y_k^*)$	$n_x(x_h^*)$
	$n_y(y_1^*)$	$n_y(y_2^*)$	$\cdots$	$n_y(y_j^*)$	$\cdots$	$n_y(y_k^*)$	N

number of observations that assume the jth level of $Y(j = 1, 2, \ldots, J)$. Note that for any contingency table the following relationship (called marginalization) holds:

$$\sum_{i=1}^{I} n_{i+} = \sum_{j=1}^{J} n_{+j} = \sum_{i=1}^{I}\sum_{j=1}^{J} n_{ij} = n.$$

We note that given an $n \times p$ data matrix (i.e. a data matrix containing p distinct variables), it is possible to construct $p(p-1)/2$ two-way contingency tables, correspondending to all possible pairs among the p qualitative variables. However, it is usually best to generate only the contingency tables for those pairs of variables that might exhibit an interesting relationship.

3.4.1 Independence and association

In order to develop indexes to describe the relationship between qualitative variables it is necessary to first introduce the concept of statistical independence. Two variables X and Y are said to be independent, for a sample of n observations, if

$$\frac{n_{i1}}{n_{+1}} = \ldots = \frac{n_{iJ}}{n_{+J}} = \frac{n_{i+}}{n}, \forall \quad i = 1, 2, \ldots, I,$$

or, equivalently,

$$\frac{n_{1j}}{n_{1+}} = \ldots = \frac{n_{Ij}}{n_{I+}} = \frac{n_{+j}}{n}, \forall \quad j = 1, 2, \ldots, J.$$

If this occurs it means that, with reference to the first equation, the (bivariate) joint analysis of the two variables X and Y does not given any additional knowledge about X than can be gained from the univariate analysis of the variable X; the same is true for the variable Y in the second equation. When this happens Y and X are said to be statistically independent. Note that the concept of statistical independence is symmetric: if X is independent of Y then Y is independent of X.

The previous conditions can be equivalently, and more conveniently, expressed as function of the marginal frequencies n_{i+} and n_{+j}. In this case X and Y are independent if

$$n_{ij} = \frac{n_{i+}n_{+j}}{n}, \quad \forall i = 1, 2, \ldots, I; \quad \forall j = 1, 2, \ldots, J.$$

In terms of relative frequencies this is equivalent to

$$p_{XY}(x_i, y_j) = p_X(x_i)p_Y(y_j), \quad \text{for every } i \text{ and for every } j.$$

When working with real data the statistical independence condition is almost never satisfied exactly; in other words, real data often show some degree of dependence among the variables.

We note that the notion of statistical independence applies to both qualitative and quantitative variables. On the other hand, measures of dependence are defined differently depending on whether the variables are quantitative or qualitative. In the first case it is possible to calculate summary measures (called correlation measures) that work both on the levels and on the frequencies. In the second case the summary measures (called association measures) must depend on the frequencies, since the levels are not metric.

For the case of quantitative variables an important relationship holds between statistical independence and the absence of correlation. If two variables X and Y are statistically independent then also $\text{Cov}(X, Y) = 0$ and $r(X, Y) = 0$. The converse is not necessarily true: two variables may be such that $r(X, Y) = 0$, even though they are not independent. In other words, the absence of correlation does not imply statistical independence.

The study of association is more complicated than the study of correlation because there are a multitude of association measures. Here we examine three different classes of these: distance measures, dependency measures, and model-based measures.

3.4.2   Distance measures

As already remarked, independence between two variables, X and Y, holds when

$$n_{ij} = \frac{n_{i+}n_{+j}}{n}, \quad \forall i = 1, 2, \ldots, I; \quad \forall j = 1, 2, \ldots, J.$$

for all joint frequencies of the contingency table. One way to provide a summary measure of the association between two variables is based on the calculation of a 'global' measure of disagreement between the frequencies actually observed (n_{ij}) and those expected under the assumption of independence between the two variables ($n_{i.}n_{.j}/n$). The original statistic proposed by Karl Pearson is the most widely used measure for assessing the hypothesis of independence between X and Y. In the general case, such a measure is defined by

$$X^2 = \sum_{i=1}^{I}\sum_{j=1}^{J} \frac{(n_{ij} - n_{ij}^*)^2}{n_{ij}^*},$$

where

$$n_{ij}^* = \frac{n_{i+}n_{+j}}{n}, \quad i = 1, 2, \ldots, I; j = 1, 2, \ldots, J.$$

Note that $X^2 = 0$ if the X and Y variables are independent. In fact in such a case, the factors in the numerator are all zero.

We note that the X^2 statistic can be written in the equivalent form

$$X^2 = n \left[\sum_{i=1}^{I}\sum_{j=1}^{J} \frac{n_{ij}^2}{n_{i+}n_{+j}} - 1 \right]$$

which emphasizes the dependence of the statistic on the number of observations, n; this is a potential problem since the value of X^2 increases with the sample size n. To overcome this problem, alternative measures have been proposed that are function of the previous statistic.

A first measure is

$$\phi^2 = \frac{X^2}{n} = \sum_{i=1}^{I} \sum_{j=1}^{J} \frac{n_{ij}^2}{n_{i+} n_{+j}} - 1,$$

usually called the mean contingency. The square root of ϕ^2 is instead called phi coefficient. Note that, in the case of a 2×2 contingency table representing binary variables, the ϕ^2 coefficient is normalised as it takes values between 0 and 1 and, furthermore, it can be shown that

$$\phi^2 = \frac{\text{Cov}^2(X, Y)}{\text{Var}(X)\text{Var}(Y)}.$$

Therefore, the ϕ^2 coefficient, in the case of 2×2 tables, is equivalent to the squared linear correlation coefficient.

However, in the case of contingency tables larger than 2×2, the ϕ^2 index is not normalised. The Cramer index normalises the X^2 measure, so that it can be used for making comparisons. The Cramer index is obtained by dividing X^2 by the maximum value it can assume for a given contingency table; this is a common approach used in descriptive statistics for normalising measures. Since such maximum can be shown to be the smaller of $I - 1$ and $J - 1$, where I and J are the number of rows and columns of the contingency table respectively, the Cramer index is equal to

$$V^2 = \frac{X^2}{n \min[I - 1, J - 1]}.$$

It can be shown that $0 \le V^2 \le 1$ for any $I \times J$ contingency table and in particular, $V^2 = 0$ if and only if X and Y are independent. On the other hand, $V^2 = 1$ in case of maximum dependency between the two variables. V^2 takes value 1 in three instances:

(a)   There is maximum dependency of Y on X when in every row of the table there is only one non-zero frequency. This happens when every level of X corresponds to one and only one level of Y. If this holds, then $V^2 = 1$ and $I \ge J$.

(b)   There is maximum dependency of X on Y when in every column of the table there is only one non-zero frequency. This means that every level of Y corresponds to one and only one level of X. This condition occurs when $V^2 = 1$ and $J \ge I$.

(c)   If both of the two previous conditions are simultaneously satisfied, that is, if $I = J$, when $V^2 = 1$ the two variables are maximally dependent.

In our exposition we have referred to the case of two-way contingency tables, involving two variables, with an arbitrary number of levels. However, the measures presented in this subsection can be easily applied to multi-way tables, extending the number of summands in the definition of X^2, to account for all table cells.

In conclusion, the association indexes based on the X^2 Pearson statistic measure the distance between the relationship of X and Y and the case of independence. They represent a generic notion of association, in the sense that they measure exclusively the distance from the independence situation, without informing on the nature of the relationship between X and Y. On the other hand, these indexes are rather general, as they can be applied in the same fashion to all kinds of contingency tables. Furthermore, the X^2 statistic has an asymptotic probabilistic (theoretical) distribution and, therefore, can also be used to assess an inferential threshold to evaluate inductively whether the examined variables are significantly dependent.

3.4.3 Dependency measures

The measures of association seen so far are all functions of the X^2 statistics and thus have the disadvantage of being hard to interpret in the majority of real applications. This important point was underlined by Goodman and Kruskal (1979), who proposed an alternative approach for measuring the association in a contingency table. The set-up followed by Goodman and Kruskal is based on the definition of indexes suited for the specific investigation context in which they are applied. In other words, such indexes are characterised by an operational meaning that defines the nature of the dependency between the available variables.

We now examine two such measures. Suppose that, in a two-way contingency table, Y is the 'dependent' variable and X the 'explanatory' variable. It is of interest to evaluate whether, for a generic observation, knowing the category of X can reduce the uncertainty as to what the corresponding category of Y might be. The 'degree of uncertainty' as to the category of a qualitative variable is usually expressed via a heterogeneity index.

Let $\delta(Y)$ indicate a heterogeneity measure for the marginal distribution of Y, expressed by the vector of marginal relative frequencies, $\{f_{+1}, f_{+2}, \ldots, f_{+J}\}$. Similarly, let $\delta(Y|i)$ be the same measure calculated on the distribution of Y conditional on the ith row of the variable X of the contingency table, $\{f_{1|i}, f_{2|i}, \ldots, f_{J|i}\}$.

An association index based on the 'proportional reduction in the heterogeneity' (error proportional reduction index, EPR), is then given (see for instance, Agresti, 1990) by

$$\mathrm{EPR} = \frac{\delta(Y) - M[\delta(Y|X)]}{\delta(Y)},$$

where $M[\delta(Y|X)]$ is the mean heterogeneity calculated with respect to the distribution of X, namely

$$M[\delta(Y|X)] = \sum_i f_{i.}\delta(Y|i),$$

with

$$f_{i\cdot} = n_{i+}/n (i = 1, 2, \ldots, I).$$

The above index measures the proportion of heterogeneity of Y (calculated through δ) that can be 'explained' by the relationship with X.

Depending on the choice of the heterogeneity index δ, different association measures can be obtained. Usually, the choice is between the Gini index and the entropy index. In the first case it can be shown that the EPR index gives rise to the so-called concentration coefficient, $\tau_{Y|X}$:

$$\tau_{Y|X} = \frac{\sum \sum f_{ij}^2 / f_{i+} - \sum f_{+j}^2}{1 - \sum_j f_{+j}^2}.$$

In the second case, using the entropy index in the ERP expression, we obtain the so-called uncertainty coefficient, $U_{Y|X}$:

$$U_{Y|X} = -\frac{\sum_i \sum_j f_{ij} \log(f_{ij} / f_{i+} \cdot f_{+j})}{\sum_j f_{+j} \log f_{+j}},$$

where, in the case of null frequencies, by convention $\log 0 = 0$. It can be shown that both $\tau_{Y|X}$ and $U_{Y|X}$ take values in the $[0,1]$ interval. Note, in particular, that:

$\tau_{Y|X} = U_{Y|X}$ if and only if the variables are independent;
$\tau_{Y|X} = U_{Y|X} = 1$ if and only if Y has maximum dependence on X.

The indexes described have a simple operational interpretation regarding specific aspects of the dependence link between the variables. In particular, both $\tau_{Y|X}$ and $U_{Y|X}$ represent alternative quantifications of the reduction of the Y heterogeneity that can be explained through the dependence of Y on X. From this viewpoint they are, in comparison to the distance measures of associations, rather specific.

On the other hand, they are less general than the distance measures. Their application requires the identification of a causal link from one variable (explanatory) to another (dependent), while the X^2-based indexes are symmetric. Furthermore, the previous indexes cannot easily be extended to contingency tables with more than two variables, and cannot be used to derive an inferential threshold.

3.4.4 Model-based measures

The last set of association measures that we present is different from the previous two sets in the that it does not depend on the marginal distributions of the variables. For ease of notation, we will assume a probability model in which cell relative frequencies are replaced by cell probabilities. The cell probabilities can be interpreted as relative frequencies as the sample size tends to infinity, therefore they have the same properties as relative frequencies.

Consider a 2×2 contingency table summarising the joint distribution of the variables X and Y; the rows report the values of X $(X = 0,1)$ and the columns the values of Y $(Y = 0,1)$. Let π_{11}, π_{00}, π_{10} and π_{01} denote the probability that an observation is classified in one of the four cells of the table. The odds ratio is a measure of association that constitutes a fundamental parameter in the statistical models for the analysis of qualitative data. Let $\pi_{1|1}$ and $\pi_{0|1}$ denote the conditional probabilities of having a 1 (a success) and a 0 (a failure) in row 1, and $\pi_{1|0}$ and $\pi_{0|0}$ the same probabilities for row 0. The odds of success for row 1 are defined by

$$\text{odds}_1 = \frac{\pi_{1|1}}{\pi_{0|1}} = \frac{P(Y = 1|X = 1)}{P(Y = 0|X = 1)},$$

and for row 0 by

$$\text{odds}_0 = \frac{\pi_{1|0}}{\pi_{0|0}} = \frac{P(Y = 1|X = 0)}{P(Y = 0|X = 0)}.$$

The odds are always non-negative, with a value greater than 1 when a success (level 1) is more probable than a failure (level 0), that is, when $P(Y = 1|X = 1) > P(Y = 0|X = 1)$. For example, if the odds equal 4 this means that a success is four times more probable than a failure. In other words, one expects to observe four successes for every failure (i.e. four successes in five events). Conversely, if the are odds are $1/4 = 0.25$ then a failure is four times more probable than a success, and one expects to observe one success for every four failures (i.e. one success in five events).

The ratio between the above two odds values is called the odds ratio:

$$\theta = \frac{\text{odds}_1}{\text{odds}_0} = \frac{\pi_{1|1}/\pi_{0|1}}{\pi_{1|0}/\pi_{0|0}}.$$

From the definition of the odds, and using the definition of joint probability, it can easily be shown that:

$$\theta = \frac{\pi_{11} \cdot \pi_{00}}{\pi_{10} \cdot \pi_{01}}.$$

This expression shows that the odds ratio is a cross product ratio, the product of probabilities on the main diagonal divided by the product of the probabilities off the main diagonal of a contingency table.

In the actual computation of the odds ratio, the probabilities will be replaced with the observed frequencies, leading to the expression

$$\theta_{ij} = \frac{n_{11}n_{00}}{n_{10}n_{01}}.$$

We now list some properties of the odds ratio, without proof.

1. The odds ratio can be equal to any non-negative number, that is, it can take values in the interval $[0, +\infty)$.
2. When X and Y are independent $\pi_{1|1} = \pi_{1|0}$, so that $\text{odds}_1 = \text{odds}_0$ and $\theta = 1$. On the other hand, depending on whether the odds ratio is greater or smaller than 1 it is possible to evaluate the sign of the association:

- for $\theta > 1$ there is a positive association, since the odds of success are greater in row 1 than in row 0;
- for $0 < \theta < 1$ there is a negative association, since the odds of success are greater in row 0 that in row 1.
3. When the order of the rows or the order of the columns is reversed, the new value of θ is the reciprocal of the original value. On the other hand, the odds ratio does not change value when the orientation of the table is reversed so that the rows become columns and the columns become rows. This means that the odds ratio deals with the variables symmetrically and, therefore, it is not necessary to identify one variable as dependent and the other as explanatory.

The odds ratio can be used as an exploratory tool aimed at building a probabilistic model, similarly to the linear correlation coefficient.

Concerning the construction of a decision rule that determines whether a certain observed value of the odds ratio indicates a significant association between the corresponding variables, it is possible to derive a confidence interval, as was done for the correlation coefficient. The interval leads to a rule for detecting a significant association when

$$|\log \theta_{ij}| > z_{\alpha/2} \sqrt{\sum_{ij} \frac{1}{\sqrt{n_{ij}}}},$$

where $z_{\alpha/2}$ is the $100(1 - \alpha/2)\%$ percentile of a standard normal distribution. For instance, when $\alpha = 0.05$, $z_{\alpha/2} = 1.96$. We remark that the confidence interval used in this case is only approximate, but that the approximation improves with the sample size.

So far we have defined the odds ratio for 2×2 contingency tables. However, odds ratios can be calculated in a similar fashion for larger contingency tables. The odds ratio for $I \times J$ tables can be defined with reference to each of the $\binom{I}{2} = I(I-2)/2$ pairs of rows in combination with each of the $\binom{J}{2} = J(J-2)/2$ pairs of columns. There are $\binom{I}{2}\binom{J}{2}$ odds ratios of this type. As the number of odds ratios to be calculated can become enormous, it is wise to choose parsimonious representations.

3.5   Reduction of dimensionality

In the analysis of complex multivariate data sets, it is often necessary to reduce the dimensionality of the problem, expressed by the number of variables present. For example, it is impossible to visualise graphs for a dimension greater than 3. A technique that is typically used to achieve this task is the linear operation known as principal components transformation. It must be emphasised that this can be used only for quantitative variables or, possibly, for binary variables. However, in practice, it is often applied to labelled qualitative data for exploratory purposes as well. In any case, the method constitutes an important reference for all dimensionality reduction techniques.

The underlying idea of the method is to transform p statistical variables (usually correlated) in terms of $k < p$ uncorrelated linear combinations, organised according to the explained variability. Consider a matrix of data $\mathbf{X}$, with n rows and p columns. The starting point of the analysis is the variance–covariance matrix, $\mathbf{S} = n^{-1}\tilde{\mathbf{X}}'\tilde{\mathbf{X}}$ (see Table 3.5). In order to simplify the notation, in the rest of this section it will be assumed that the observations are already expressed in terms of deviations from the mean and, therefore, $\mathbf{X} = \tilde{\mathbf{X}}$. We remark that, whenever the variables are expressed according to different measurement scales, it is best to standardise all the variables before calculating $\mathbf{S}$. Alternatively, it is sufficient to substitute $\mathbf{S}$ with the correlation matrix $\mathbf{R}$, since $\mathbf{R} = n^{-1}\mathbf{Z}'\mathbf{Z}$ (see Table 3.6). In any case, it is assumed that both $\mathbf{S}$ and $\mathbf{R}$ are of full rank; this implies that none of the variables considered is a perfect linear function of the others (or a linear combination of them).

The computational algorithm for principal components can be described in an iterative way. Note that in this section the symbols that represent vectors are underlined (this is the conventional notation in linear algebra), so that they can be distinguished from matrices (indicated with capital letters) and from scalar constants (denoted by a standard character).

Definition. The *first principal component* of the data matrix $\mathbf{X}$ is a vector described by the following linear combination of the variables:

$$\begin{pmatrix} Y_{11} \\ \vdots \\ Y_{n1} \end{pmatrix} = a_{11} \begin{pmatrix} x_{11} \\ \vdots \\ x_{n1} \end{pmatrix} + a_{21} \begin{pmatrix} x_{12} \\ \vdots \\ x_{n2} \end{pmatrix} + \cdots + a_{p1} \begin{pmatrix} x_{1p} \\ \vdots \\ x_{np} \end{pmatrix},$$

that is, in matrix terms,

$$\mathbf{Y}_1 = \sum_{j=1}^{p} a_{j1}\mathbf{X}_j = \mathbf{X}\mathbf{a}_1.$$

Furthermore, in the previous expression, the vector of the coefficients (also called weights) $\mathbf{a}_1 = (a_{11}, a_{21}, \ldots, a_{p1})'$ is chosen to maximise the variance of the variable $\mathbf{Y}_1$. In order to obtain a unique solution it is required that the weights are normalised, constraining the sum of their squares to be 1. Therefore, the first principal component is determined by the vector of weights a_1 such that max $\mathrm{Var}(\mathbf{Y}_1) = \max(\mathbf{a}_1, \mathbf{S}\mathbf{a}_1)$, under the constraint $\mathbf{a}'_1\mathbf{a}_1 = 1$, which normalises the vector.

The solution of the previous problem is obtained using Lagrange multipliers. It can be shown that, in order to maximise the variance of $\mathbf{Y}_1$, the weights can be chosen to be the eigenvector corresponding to the largest eigenvalue of the variance–covariance matrix $\mathbf{S}$. We omit the proof, which can be found in a multivariate statistic text, such as Mardia *et al.* (1979).

Definition. The second principal component of the data matrix $\mathbf{X}$ is the linear combination:

$$\begin{pmatrix} Y_{12} \\ \vdots \\ Y_{n2} \end{pmatrix} = a_{12} \begin{pmatrix} x_{11} \\ \vdots \\ x_{n1} \end{pmatrix} + a_{22} \begin{pmatrix} x_{12} \\ \vdots \\ x_{n2} \end{pmatrix} + \cdots + a_{p2} \begin{pmatrix} x_{1p} \\ \vdots \\ x_{np} \end{pmatrix},$$

that is, in matrix terms,

$$\mathbf{Y}_2 = \sum_{j=1}^{p} a_{j2} \mathbf{X}_j = \mathbf{X} \mathbf{a}_2,$$

where the vector of the coefficients $\mathbf{a}_2 = (a_{12}, \ldots, a_{p2})'$ is chosen in such a way that max $\mathrm{Var}(\mathbf{Y}_2) = \max(\mathbf{a}_2, \mathbf{S}\mathbf{a}_2)$, under the constraints $\mathbf{a}'_2 \mathbf{a}_2 = 1$ and $\mathbf{a}'_2 \mathbf{a}_1 = 0$. Note the second constraint, which requires the two vectors $\mathbf{a}_2$ and $\mathbf{a}_1$ orthogonal. This means that the first and second components will be uncorrelated. The expression for the second principal component can be obtained through the method of Lagrange multipliers, and $\mathbf{a}_2$ is the eigenvector (normalised and orthogonal to $\mathbf{a}_1$) corresponding to the second largest eigenvalue of $\mathbf{S}$.

This process can be used recursively to define the kth principal component, with k less than the number of variables p. In general, the vth principal component, for $v = 1, \ldots, k$, is given by the linear combination

$$\mathbf{Y}_v = \sum_{j=1}^{p} a_{jv} \mathbf{X}_j = \mathbf{X} \mathbf{a}_v$$

in which the vector of the coefficients $\mathbf{a}_v$ is the eigenvector of $\mathbf{S}$ corresponding to the vth largest eigenvalue. This eigenvector is normalised and orthogonal to all the previous eigenvectors.

3.5.1 Interpretation of the principal components

The main difficulty with the principal components is their interpretation. This is because each principal component is a linear combination of all the available variables, hence they do not have a clear measurement scale. To facilitate their interpretation, we will now introduce the concepts of absolute and relative importance of the principal components.

We begin with the absolute importance. To solve the maximisation problem that leads the principal components, it can be shown that $\mathbf{S}\mathbf{a}_v = \lambda_v \mathbf{a}_v$. Therefore, the variance of the vth principal component corresponds to the vth eigenvalue of the data matrix:

$$\mathrm{Var}(\mathbf{Y}_v) = \mathrm{Var}(\mathbf{X}\mathbf{a}_v) = \mathbf{a}'_v \mathbf{S} \mathbf{a}_v = \lambda_v.$$

The covariance between the principal components satisfies

$$\text{Cov}(\mathbf{Y}_i, \mathbf{Y}_j) = \text{Cov}(\mathbf{Xa}_i, \mathbf{Xa}_j) = \mathbf{a}'_i \mathbf{Sa}_j = \mathbf{a}'_i \lambda_j \mathbf{a}_j = 0,$$

because $\mathbf{a}_i$ and $\mathbf{a}_j$ are assumed to be orthogonal. This implies that the principal components are uncorrelated. The variance–covariance matrix between them is thus expressed by the diagonal matrix

$$\text{Var}(Y) = \begin{bmatrix} \lambda_1 & & 0 \\ & \ddots & \\ 0 & & \lambda_k \end{bmatrix}.$$

Consequently, the following ratio expresses the proportion of variability that is 'maintained' in the transformation from the original p variables to $k < p$ principal components:

$$\frac{\text{tr}(\text{Var}Y)}{\text{tr}(\text{Var}X)} = \frac{\sum_{i=1}^{k} \lambda_i}{\sum_{i=1}^{p} \lambda_i}.$$

This equation expresses a cumulative measure of the quota of variability (and therefore of the statistical information) 'reproduced' by the first k components, with respect to the overall variability present in the original data matrix, as measured by the trace of the variance–covariance matrix. Therefore, it can be used as a measure of absolute importance of the chosen k principal components, in terms of 'quantity of information' maintained by going from p variables to k components.

We now examine the relative importance of each principal component (with respect to the single original variables). To achieve this aim we first obtain the general expression for the linear correlation between a principal component and an original variable. We have that

$$\text{Cov}(Y_j, X) = \text{Cov}(\mathbf{Xa}_j, X) = \mathbf{Sa}_j = \lambda_j \mathbf{a}_j$$

and, therefore, $\text{Cov}(Y_j, X_i) = \lambda_j a_{ij}$. Furthermore, writing s_i^2 for $\text{Var}(X_i)$ and recalling that $\text{Var}(Y_v) = \lambda_v$, we have that

$$\text{Corr}(Y_j, X_i) = \frac{\sqrt{\lambda_j} a_{ji}}{s_i}.$$

Notice that the algebraic sign and the value of the coefficient (called also loading) a_{ji}, determine the sign and the strength of the correlation between the jth component and the ith original variable. It also follows that the portion of variability of an original variable, say X_i, explained by k principal components can be described by the expression

$$\sum_{j=1}^{k} \text{Corr}^2(Y_j, X_i) = \frac{\lambda_1 a_{1i}^2 + \cdots + \lambda_k a_{ki}^2}{s_i^2},$$

which describes the quota of variability (information) of each explanatory variable that is maintained in going from the original variables to the principal components. This permits us to interpret each principal component talking about the variables with which it is mostly correlated (in absolute value).

We conclude this subsection with three remarks on principal components analysis:

- The method of principal components permit us to reduce the complexity of a data matrix, in terms of number of variables, going from a data matrix $\mathbf{X}_{n \cdot p}$ to a matrix with fewer columns, according to the transformation $\mathbf{Y}_{n \cdot k} = \mathbf{X}_{n \cdot p} \mathbf{A}_{p \cdot k}$, where $\mathbf{A}_{p \cdot k}$ is the matrix obtained stacking columnwise the eigenvectors corresponding to the principal components. The resulting transformed observations are usually called principal components scores and 'reproduce' the data matrix in a space of lower dimension.

- The principal components can be calculated by extracting the eigenvalues and the corresponding eigenvectors from the correlation matrix $\mathbf{R}$ instead than from the variance–covariance matrix $\mathbf{S}$. The principal components obtained from $\mathbf{R}$ are not the same as those obtained from $\mathbf{S}$. In order to choose which matrix to start from, in general, use $\mathbf{R}$ when the variables are expressed in different measurement scales. Note also that, using $\mathbf{R}$, the interpretation of the importance of components is simpler. In fact, since the $\text{tr}(\mathbf{R}) = p$, the degree of absolute importance of k components is given by:

$$\frac{\text{tr}(\text{Var} Y)}{\text{tr}(\text{Var} X)} = \frac{\sum_{i=1}^{k} \lambda_i}{p}$$

while the degree of relative importance of a principal component, with respect to a variable, is

$$\text{Corr}(Y_j, X_i) = \sqrt{\lambda_i} a_{ji}.$$

- How many principal components should we choose? This is a critical point, for which there are different empirical criteria. One solution involves considering all the components that have an absolute degree of importance larger than a certain threshold thought to be reasonable, such as 50%. Or else, if $\mathbf{R}$ has been used, it is possible to choose all the principal components with corresponding eigenvalues greater than 1; since the overall variance equals p, the average variance of the components should be approximately equal to 1. Finally, a graphical instrument that is quite useful for determining the number of components is the so called 'scree plot', which plots on the x-axis the index of the component $(1, 2, 3, \ldots, k)$, and on the y-axis the corresponding eigenvalue. An empirical rule suggests choosing, as number of components, the value corresponding to the point where there is a significant 'fall' in the y-axis.

As alternatives to the empirical criteria here presented, there are inferential type criteria that require the assumption of a specific probability model; for more details, see Mardia et al. (1979).

3.6 Further reading

Exploratory analysis has developed as an autonomous field of statistics, in parallel with the development of the computing resources. It is possible to date the initial developments in the field to the publication of the texts by Benzécri (1973) and Tukey (1977).

Having briefly described the main analogies and differences between data mining and exploratory analysis, in this chapter we described the main exploratory data analysis methods. We began by focusing on univariate exploratory analysis. This phase is often fundamental to understanding what might be discovered during a data mining analysis. It often reveals problems with data quality, such as missing items and anomalous data.

Since the observed reality is typically multidimensional, the next phase in exploratory analysis is multivariate in nature. Given the difficulty in visualising multidimensional phenomena, many analyses focus on bivariate exploratory analysis, and on how the relationships found in a bivariate analysis can modify themselves, conditioning the analysis on the other variables. Similar considerations apply to qualitative variables, for example comparing the marginal odds ratios with those calculated conditionally. In the latter case a phenomenon known as Simpson's paradox (see, for example, Agresti, 1990) is observed, for which a certain observed marginal association can completely change direction when conditioning the odds ratio on the level of additional variables.

We focused on some important matrix representations which are of use when conducting a more comprehensive multidimensional exploratory analysis of the data. We refer the reader interested in the use of matrix calculations in statistics to Searle (1982). Multidimensional exploratory data analysis remains an active area of research in statistics thanks to developments in computer science. We expect, therefore, that there will be substantial advances in this research area in the near future. For a review of some of these developments, particularly multidimensional graphics, see Hand *et al.* (2001) or, from a computational statistics viewpoint, the text of Venables and Ripley (2002).

We then introduced the multidimensional analysis of qualitative data. This topic also remains an active research area, and the existence of a large number of indexes shows that the subject has yet to be consolidated. We put the available indexes into three principal classes: distance measures, dependence measures and model-based indexes. Distance measures are applicable to any contingency table. Dependence measures, in contrast, give precise information on the type of dependence among the variables under examination, but are hardly applicable to contingency tables of dimension greater than 2. Model-based indexes are a possible compromise. They are sufficiently broad and offer a good amount of information. In addition, they have the advantage of characterizing the most important statistical models for the analysis of qualitative data: the logistic and loglinear regression models. For an introduction to the descriptive analysis of qualitative data, see Agresti (1990).

An alternative approach to the multidimensional visualization of data is the reduction to the principal components. The method has been described from an applied point of view; for more details on the formal aspects of this method, see Mardia *et al.* (1979). The method of principal components has a very important role in factor analysis. This method assumes a probability model, usually Gaussian. It decomposes the variance–covariance matrix into two parts, one part common to all the variables corresponding to the presence of underlying latent (unobserved or unmeasurable) variables, and the other part specific to each variable. In this framework, the chosen principal components identify the latent variables and are interpretated accordingly. In addition, it is possible to employ methods of 'rotation' of the components (latent factors) that modify the weight coefficients, improving the interpretability. For further details on factor analysis we refer the reader to Bollen (1989).

Principal components analysis is probably one of the simplest data reduction methodsas it is based on linear transformations. Essentially, the scores obtained transform the original data into linear projections on the reduced space, minimising the Euclidean distance between the coordinates in the original space and the transformed data. Other types of transformations include wavelet methods, based on Fourier transforms, as well as the methods of projection pursuit, which look for the best directions of projection on a reduced space. For both techniques we refer the reader to other data mining texts, such as Hand *et al.* (2001) or Hastie *et al.* (2001).

Methods of data or dimension reduction are also available for qualitative data. For a contingency table with two dimensions, correspondence analysis produces a row (column) profile for every row (column), corresponding to the conditional frequency distribution of the row (column). Dimension reduction is then performed by projecting such profiles in a space of lower dimension that reproduces as much of the original inertia as possible, the latter being related to the X^2 statistics. Correspondence analysis can also be applied to contingency tables of arbitrary dimension (represented through the so-called Burt matrix). For an introduction to correspondence analysis, see, for instance, Greenacre (1983).

CHAPTER 4

Model specification

This chapter introduces the main data mining methods. It is appropriate to divide these into two main groups. The first group (Sections 4.1–4.8) consists of methods that do not necessarily require the specification of a probability model. In fact, many of these methods were developed by computer scientists rather than statisticians. Recently however, statisticians have adopted these methods because of their effectiveness in solving data mining problems.

For the second group of methods (Sections 4.9–4.15) it is essential to adopt a probability model which describes the data generating mechanism. The introduction of such a framework allows more subtle information to be extracted from the data; on the other hand, it requires more assumptions to be made. Most of the methods belonging to this second group were developed by statisticians. However, they have also been adopted by computer scientists working in data mining, because of their greater accuracy.

Section 4.1 deals with the important concepts of proximity and distance between statistical observations, which is the foundation for many of the methods discussed in the chapter. Section 4.2 deals with clustering methods, the aim of which is to classify observations into homogeneous groups. Clustering is probably the best known descriptive data mining method. In Section 4.3 we present linear regression from a non-probabilistic viewpoint. This is the most important prediction method for continuous variables. We will present the probability aspects of linear regression in Section 4.11. In Section 4.4 we examine, again from a non-probabilistic viewpoint, the main prediction method for qualitative variables: logistic regression. Another important predictive methodology is represented by tree models, which can be used both for regression and clustering purposes. These are presented in Section 4.5. Concerning clustering, there is a fundamental difference between cluster analysis, on the one hand, and logistic regression and tree models, on the other. In the latter case, the clustering is supervised, that is, measured against a reference variable (target or response), whose values are known. The former case, in contrast, is unsupervised: there are no reference variables, and the clustering analysis determines the nature and the number of groups and allocates the observations in them. In Sections 4.6 and 4.7 we introduce two further classes of predictive models, neural networks and nearest-neighbour models. Then in Section 4.8 we describe two very important local data mining methods: association and sequence rules.

Applied Data Mining for Business and Industry, 2e   P. Giudici, S. Figini
© 2009 John Wiley & Sons, Ltd

In Section 4.9 we introduce data mining methods that require the specification of an underlying probability model. Such methods allow more powerful and more interpretable results to be derived, since they can make use of concepts of statistical inference. We start the section with an introduction to the measurement of uncertainty using probability and to the basic concepts of statistical inference. In particular, we introduce the Gaussian distribution, the most popular parametric probability model. We then move, in Section 4.10, to non-parametric and semiparametric modelling of the data, and show how these approaches can be used. We introduce a probability approach to cluster analysis, based on mixture models, as well as the basic ideas behind kernel density estimation.

Section 4.11 introduces the normal linear model as the main tool in modelling the relationship between one or more response variables and one or more explanatory variables, with the aim of constructing a decision rule which permits us to predict the values of the response variables, given the values of the explanatory variables.

In Section 4.12 we introduce a more general class of parametric models, based on the exponential family of distributions, and thus derive a more general class of linear models (called generalised linear models), that contains, as special cases, the linear model and the logistic regression model. Another important class of generalised linear models are the log-linear models, introduced in Section 4.13, which constitute the most important data mining tool for descriptively analysing the relationships between qualitative variables. In Section 4.14 we extend the logic of this modelling with graphical models.

Finally, in Section 4.15 we introduce survival analysis models, originally developed for medical applications, but now increasingly used in the business field as well.

4.1 Measures of distance

In this chapter we will often discuss methods suitable for classifying and grouping observations into homogeneous groups. In other words, we will consider the relationships between the rows of the data matrix which correspond to observations. In order to compare observations, we need to introduce the idea of a distance measure, or proximity, among them. The indexes of proximity between pairs of observations furnish indispensable preliminary information for identifying homogeneous groups. More precisely, an index of proximity between any two observations x_i and x_j can be defined as a function of the corresponding row vectors in the data matrix:

$$IP_{ij} = f(x_i', x_j'), \quad i, j = 1, 2, \ldots, n.$$

We will use an example from Chapter 6 as a running example in this section. We have $n = 32711$ visitors to a website and $p = 35$ dichotomous variables that define the behaviour of each visitor. In this case, a proximity index will be a function of two 35-dimensional row vectors. Knowledge of the indexes of proximity for every pair of visitors allows us to select those among them who are more similar, or at least less different, with the purpose of identifying some groups as the most homogeneous among them.

When the variables of interest are quantitative, the indexes of proximity typically used are called distances. If the variables are qualitative, the distance between observations can be measured by indexes of similarity. If the data are contained in a contingency table, the chi-squared distance can also be employed. There are also indexes of proximity that are used on a mixture of qualitative and quantitative variables. We will examine the Euclidean distance for quantitative variables, and some indexes of similarity for qualitative variables.

4.1.1 Euclidean distance

Consider a data matrix containing only quantitative (or binary) variables. If x and y are rows from the data matrix then a function $d(x, y)$ is said to be a distance between two observations if it satisfies the following properties:

- **Non-negativity.** $d(x, y) \geq 0$, for all x and y.
- **Identity.** $d(x, y) = 0 \Leftrightarrow x = y$, for all x and y.
- **Symmetry.** $d(x, y) = d(y, x)$, for all x and y.
- **Triangular inequality.** $d(x, y) \leq (x, z) + d(y, z)$, for all x, y and z.

To achieve a grouping of all observations, the distance is usually considered between all observations present in the data matrix. All such distances can be represented in a matrix of distances. A distance matrix can be represented in the following way:

$$\Delta = \begin{pmatrix} 0 & \cdots & d_{1i} & \cdots & d_{1n} \\ \vdots & \ddots & \vdots & & \vdots \\ d_{i1} & \cdots & 0 & \cdots & d_{in} \\ \vdots & & \vdots & \ddots & \vdots \\ d_{n1} & \cdots & d_{ni} & \cdots & 0 \end{pmatrix},$$

where the generic element d_{ij} is a measure of distance between the row vectors x_i and x_j. The Euclidean distance is the most commonly used distance measure. It is defined, for any two units indexed by i and j, as the square root of the difference between the corresponding vectors, in the p-dimensional Euclidean space:

$$_2 d_{ij} = d(x_i, x_j) = \left[\sum_{s=1}^{p} (x_{is} - x_{js})^2 \right]^{1/2}.$$

The Euclidean distance can be strongly influenced by a single large difference in one dimension of the values, because the square will greatly magnify that difference. Dimensions having different scales (e.g. some values measured in centimetres, others in metres) are often the source of these overstated differences. To overcome such limitation, the Euclidean distance is often calculated, not on the original variables, but on useful transformations of them. The most common choice is to standardise the variables. After standardisation, every transformed variable contributes to the calculation of the distance with equal weight. When the variables are standardised, they have zero mean and unit variance; furthermore, it can be shown that, for $i, j = 1, \ldots, p$:

$$_2d_{ij}^2 = 2(1 - r_{ij}),$$

$$r_{ij} = 1 - d_{ij}^2/2,$$

where r_{ij} is the correlation coefficient between the observations x_i and x_j. Thus the Euclidean distance between two observations is a function of the correlation coefficient between them.

4.1.2   Similarity measures

Given a finite set of observations $u_i \in U$, a function $S(u_i, u_j) = S_{ij}$ from $U \times U$ to $\mathbb{R}$ is called an index of similarity if it satisfies the following properties:

- **Non-negativity.** $S_{ij} \geq 0$, for all $u_i, u_j \in U$.
- **Normalisation.** $S_{ii} = 1$, for all $u_i \in U$.
- **Symmetry.** $S_{ij} = S_{ji}$, for all $u_i, u_j \in U$.

Unlike distances, the indexes of similarity can be applied to all kinds of variables, including qualitative variables. They are defined with reference to the observation indexes, rather than to the corresponding row vectors, and they assume values in the closed interval [0, 1], making them easy to interpret.

The complement of an index of similarity is called an index of dissimilarity and represents a class of indexes of proximity wider than that of the distances. In fact, as a distance, a dissimilarity index satisfies the properties of non-negativity and symmetry. However, the property of normalisation is not equivalent to the property of identity of the distances. Finally, dissimilarities do not have to satisfy the triangle inequality.

As we have observed, indexes of similarity can be calculated, in principle, for quantitative variables. But they would be of limited use since they would tell us only whether two observations had, for the different variables, observed values equal or different, without saying anything about the size of the difference. From an operational viewpoint, the principal indexes of similarity make reference to data matrices containing binary variables. More general cases, with variables having more than two levels, can be brought back to this setting through the technique of binarisation.

Consider data on n visitors to a website, which has P pages. Correspondingly, there are P binary variables, which assume the value 1 if the specific page has been visited, or else the value 0. To demonstrate the application of similarity indexes, we now analyse only data concerning the behaviour of the first two visitors (2 of the n observations) to the website described in Chapter 6, among the $P = 28$ web pages that they can visit. Table 4.1 summarises the behaviour of the two visitors, treating each page as a binary variable.

Table 4.1 Classification of the visited web pages.

Visitor A \ Visitor B	1	0	Total
1	$CP = 2$	$PA = 4$	6
0	$AP = 1$	$CA = 21$	22
Total	3	25	$P = 28$

Note that, of the 28 pages considered, two have been visited by both visitors. In other words, 2 represent the absolute frequency of contemporary occurrences (CP, for co-presence, or positive matches) for the two observations. In the lower right-hand corner of the table, there is a frequency of 21 equal to the number of pages that are visited neither by A nor by B. This frequency corresponds to contemporary absences in the two observations (CA, for co-absences or negative matches). Finally, the frequencies of 4 and 1 indicate the number of pages that only one of the two navigators visits (PA for presence–absence and AP for absence–presence, where the first letter refers to visitor A and the second to visitor B).

The latter two frequencies denote the differential aspects between the two visitors and therefore must be treated in the same way, being symmetrical. The co-presence is aimed at determining the similarity among the two visitors, a fundamental condition because they could belong to the same group. The co-absence is less important, perhaps of negligible importance for determining the similarities between the two units. In fact, the indexes of similarity developed in the statistical literature differ in how they treat the co-absence, as we now describe.

Russel–Rao similarity index
The Russel–Rao similarity index is a function of the co-presences and is equal to the ratio between the number of the co-presences and the total number of binary variables considered, P:

$$S_{ij} = \frac{CP}{P}.$$

From Table 4.1 we have

$$S_{ij} = \frac{2}{28} \approx 0.07.$$

Jaccard similarity index
This index is the ratio between the number of co-presences and the total number of variables, excluding those that manifest co-absences:

$$S_{ij} = \frac{CP}{CP + PA + AP}.$$

Note that this index cannot be defined if two visitors or, more generally, the two observations, manifest only co-absences $(CA = P)$. In the example above we have

$$S_{ij} = \frac{2}{7} \approx 0.29.$$

Sokal–Michener similarity index

This is the ratio between the number of co-presences or co-absences and the total number of the variables:

$$S_{ij} = \frac{CP + CA}{P}.$$

In our example

$$S_{ij} = \frac{23}{28} \approx 0.82.$$

For the Sokal–Michener index (also called the simple matching coefficient or, in a slight abuse of terminology, the binary distance) it is simple to demonstrate that its complement to one (a dissimilarity index) corresponds to the average of the squared Euclidean distance between the two vectors of binary variables associated to the observations:

$$1 - S_{ij} = \frac{1}{P}(_2 d_{ij}^2).$$

It is one of the commonly used indexes of similarity.

4.1.3   Multidimensional scaling

In the previous subsections we have seen how to calculate proximities between observations, on the basis of a given data matrix, or a table derived from it. Sometimes, only the proximities between observations are available, for instance in terms of a distance matrix, and it is desired to reconstruct the values of the observations. In other cases, the proximities are calculated using a dissimilarity measure and it is desired to reproduce them in terms of a Euclidean distance, to obtain a representation of the observations in a two-dimensional plane. Multidimensional scaling methods are aimed at representing observations whose observed values are unknown (or not expressed numerically) in a low-dimensional Euclidean space (usually $\mathbb{R}^2$). The representation is achieved by preserving the original distances as far as possible.

Section 3.5 explained how to use the method of principal components on a quantitative data matrix in a Euclidean space. It turns the data matrix into a lower-dimensional Euclidean projection by minimising the Euclidean distance between the original observations and the projected ones. Similarly, multidimensional scaling methods look for low-dimensional Euclidean representations of the observations, representations which minimise an appropriate distance between the original distances and the new Euclidean distances. Multidimensional scaling methods differ in how such distance is defined. The most common choice is the stress function, defined by

$$\sqrt{\sum_{i=1}^{n} \sum_{j=1}^{n} (\delta_{ij} - d_{ij})^2},$$

where the δ_{ij} are the original distances (or dissimilarities) between each pair of observations, and the d_{ij} are the corresponding distances between the reproduced coordinates.

Metric multidimensional scaling methods look for k real-valued n-dimensional vectors, each representing one coordinate measurement of the n observations, such that the $n \times n$ distance matrix between the observations, expressed by d_{ij}, minimises the squared stress function. Typically, $k = 2$, so that the results of the procedure can be conveniently represented in a scatterplot. The illustrated solution is also known as least squares scaling. A variant of least squares scaling is Sammon mapping, that minimises

$$\sqrt{\sum_{i=1}^{n} \sum_{j=1}^{n} \frac{(\delta_{ij} - d_{ij})^2}{\delta_{ij}}},$$

thereby preserving smaller distances.

When the proximities between objects are expressed by a Euclidean distance, it can be shown that the solution of the previous problem corresponds to the principal component scores that would be obtained if the data matrix were available.

It is possible to define non-metric multidimensional scaling methods, where the relationship preserved between the original and the reproduced distances is not necessarily Euclidean. For further details, see Mardia *et al.* (1979).

4.2 Cluster analysis

This section is about methods for grouping a given set of observations, known as cluster analysis. Cluster analysis is the best-known descriptive data mining method. Given a data matrix composed of n observations (rows) and p variables (columns), the objective of cluster analysis is to cluster the observations into groups that are internally homogeneous (internal cohesion) and heterogeneous from group to group (external separation). Note that the constitution of homogeneous groups of observations can be interpreted as a reduction of the dimension of the space $\mathbb{R}^n$, but not in the same way as in principal components analysis (Section 3.5). In fact, in a cluster analysis, the n observations are grouped into g subsets (with $g < n$), whereas in principal components analysis the p statistical variables are transformed into k new variables (with $k < p$). There are several ways to perform a cluster analysis. It is therefore important to have a clear understanding of how the analysis will proceed. Here are some important points to consider.

Choice of variables to be used

The choice of variables to be used for clustering has to consider all the relevant aspects to achieve the stated objectives. Bear in mind that using variables of little importance will inevitably worsen the results. This is a crucial problem since it will strongly condition the final result. In general, clustering can be considered satisfactory when it does not show an excessive sensitivity to small changes in the set of variables used. Before doing a cluster analysis, it is prudent to conduct accurate exploratory investigations that are able to suggest possible final configurations for the clustering. To help with visualisation and interpretation of the results, it is often appropriate to reduce the dimensionality of the data matrix, perhaps through the method of the principal components.

During the exploratory phase, pay particular attention to anomalous observations that might negatively affect the analysis. Some data mining textbooks (e.g. Han and Kamber, 2001) link the methods of cluster analysis with those that search for outliers. Although there are similarities, we take the view that one should choose cluster analysis to classify the data into groups and outlier detection to search for anomalous observations.

Method of group formation

We can distinguish hierarchical and non-hierarchical methods. Hierarchical methods allow us to get a succession of groupings (called partitions or clusters) with a number of groups from n to 1, starting from the simplest, where all the observations are separated, to the situation where all the observations belong to a unique group. The non-hierarchical methods allow us to gather the n units directly into a number of previously defined groups.

Type of proximity index

Depending to the nature of the available variables, it is necessary to define a measure of proximity among the observations, to be used for calculating distances between them. If the variables are predominantly quantitative, use the Euclidean distance; if they are predominantly qualitative, use an index of similarity; finally, if they are available in a contingency table format, use the chi-squared distance between the levels. As shown in Section 4.1, most measures of proximity can be interpreted as distances, so we will make exclusive reference to this concept. Remember the importance of standardising the variables so that all carry the same weight in the final results.

Besides establishing a measure of proximity between observations, for hierarchical clustering methods we need to establish how to calculate the distances between groups. It is usually appropriate to use the same type of distance as the distance between observations. It remains to establish which units (or synthesis of them) to use as 'representative' of the group. This depends on the hierarchical clustering method.

Choice of evaluation criteria

Evaluating the results of the grouping obtained means verifying that the groups are consistent with the primary objective of the cluster analysis and that they therefore satisfy the conditions of internal cohesion and external separation. Choosing the number of groups is of fundamental importance. There is a trade-off between obtaining homogeneous groups, which typically increases the number of groups, and the need for a parsimonious representation, which reduces the number of groups. We will return to this point.

4.2.1 Hierarchical methods

Hierarchical clustering methods allow us to get a family of partitions, each associated with the subsequent levels of grouping among the observations, calculated on the basis of the available data. The different families of partitions can be graphically represented through a tree-like structure called a hierarchical clustering tree or dendrogram. This structure associates with every step of the hierarchical procedure, corresponding to a fixed number of groups g, one and only one clustering of the observations in the g groups.

A hierarchical clustering tree can be represented as in Figure 4.1, where for simplicity we suppose there are only five observations available, numbered from 1 to 5. The branches of the tree describe subsequent clusterings of the observations. At the root of the tree, all the observations are contained in only one class. The branches of the tree indicate divisions of the observations into clusters. The five terminal nodes indicate the situation where each observation belongs to a separate group.

Agglomerative clustering is when the groups are formed from the branches to the root (from left to right in Figure 4.1). Divisive clustering is when the groups are formed from the root to the branches. Statistical software packages usually report the whole dendrogram, from the root to a number of terminal branches equal to the number of observations. It then remains to choose the optimal number of groups. This will identify the result of the cluster analysis, since in a dendrogram the choice of the number of groups g identifies a unique

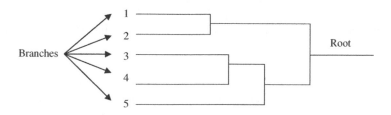

Figure 4.1 Structure of the dendrogram.

Table 4.2 Partitions corresponding to the dendrogram in Figure 4.1.

Number of clusters	Clusters
5	(1) (2) (3) (4) (5)
4	(1,2) (3) (4) (5)
3	(1,2) (3,4) (5)
2	(1,2) (3,4,5)
1	(1,2,3,4,5)

partition of the observations. For example, the partitions of the five observations described by the dendrogram in Figure 4.1 can be represented as in Table 4.2.

Table 4.2 shows that the partitions described by a dendrogram are nested. This means that, in the hierarchical methods, the elements that are united (or divided) at a certain step will remain united (separated) until the end of the clustering process. Supposing we consider an agglomerative method that proceeds from five groups to one group; then units 1 and 2 are united at the second step and remain in the same group until the end of the procedure. Nesting reduces the number of partitions to compare, making the procedure computationally more efficient, but the disadvantage is not being able 'to correct' errors of clustering committed in the preceding steps. Here is an outline for an agglomerative clustering algorithm:

1. **Initialisation.** Given n statistical observations to classify, every element represents a group (in other words, the procedure starts with n clusters). The clusters will be identified with a number that goes from 1 to n.
2. **Selection.** The two 'nearest' clusters are selected, in terms of the distance initially fixed, for example, in terms of the Euclidean distance.
3. **Updating.** The number of clusters is updated (to $n - 1$) through the union, in a unique cluster, of the two groups selected in step 2. The matrix of the distances is updated, taking the two rows (and two columns) of distances between the two clusters and replacing them with only one row (and one column) of distances, 'representative' of the new group. Different clustering methods define this representation in different ways.
4. **Repetition.** Steps 2 and 3 are performed $n - 1$ times.
5. **End.** The procedure stops when all the elements are incorporated in a unique cluster.

We will now look at some of the different clustering methods mentioned in step 3. They will be introduced with reference to two groups, C_1 and C_2. Some methods require only the distance matrix and some require the distance matrix plus the original data matrix. These examples require only the distance matrix:

- **Single linkage.** The distance between two groups is defined as the minimum of the $n_1 n_2$ distances between each observation of group C_1 and each observation of group C_2:

$$d(C_1, C_2) = \min(d_{rs}), \text{ with } r \in C_1, s \in C_2.$$

- **Complete linkage.** The distance between two groups is defined as the maximum of the $n_1 n_2$ distances between each observation of a group and each observation of the other group:

$$d(C_1, C_2) = \max(d_{rs}), \text{ with } r \in C_1, s \in C_2.$$

- **Average linkage.** The distance between two groups is defined as the arithmetic average of the $n_1 n_2$ distances between each of the observations of a group and each of the observations of the other group:

$$d(C_1, C_2) = \frac{1}{n_1 n_2} \sum_{r=1}^{n_1} \sum_{s=1}^{n_2} d_{rs}, \text{ with } r \in C_1, s \in C_2.$$

Two methods that require the data matrix as well as the distance matrix are the centroid method and Ward's method.

Centroid method

The distance between two groups C_1 and C_2, having n_l and n_2 observations respectively, is defined as the distance between the respective centroids (usually the means), $\bar{x}_1$ and $\bar{x}_2$:

$$d(C_1, C_2) = d(\bar{x}_1, \bar{x}_2).$$

To calculate the centroid of a group of observations we need the original data, and we can obtain that from the data matrix. It will be necessary to replace the distances with respect to the centroids of the two previous clusters by the distances with respect to the centroid of the new cluster. The centroid of the new cluster can be obtained from

$$\frac{\bar{x}_1 n_1 + \bar{x}_2 n_2}{n_1 + n_2}.$$

Note the similarity between this method and the average linkage method: the average linkage method considers the average of the distances among the observations of each of the two groups, while the centroid method calculates the centroid of each group and then measures the distance between the centroids.

Ward's method

In choosing the groups to be joined, Ward's method minimises an objective function using the principle that clustering aims to create groups which have maximum internal cohesion and maximum external separation.

The total deviance (T) of the p variables, corresponding to n times the trace of the variance–covariance matrix, can be divided in two parts: the deviance within the groups (W) and the deviance between the groups (B), so that $T = W + B$. This is analogous to dividing the variance into two parts for linear regression

(Section 4.3). In that case B is the variance explained by the regression and W is the residual variance, the variance not explained by the regression. In formal terms, given a partition into g groups, the total deviance of the p variables corresponds to the sum of the deviances of the single variables, with respect to the overall mean $\bar{x}_s$, defined by

$$T = \sum_{s=1}^{p} \sum_{i=1}^{n} (x_{is} - \bar{x}_s)^2.$$

The deviance within groups is given by the sum of the deviances of each group,

$$W = \sum_{k=1}^{g} W_k,$$

where W_k represents the deviance of the p variables in the kth group (number n_k and centroid $\bar{x}_k = [\bar{x}_{1k}, \ldots, \bar{x}_{pk}]'$), given by

$$W_k = \sum_{s=1}^{p} \sum_{i=1}^{n_k} (x_{is} - \bar{x}_{sk})^2.$$

The deviance between groups is given by the sum (calculated on all the variables) of the weighted deviances of the group means with respect to the corresponding general averages:

$$B = \sum_{s=1}^{p} \sum_{k=1}^{g} n_k (\bar{x}_{sk} - \bar{x}_s)^2.$$

Using Ward's method, groups are joined so that the increase in W is smaller and the increase in B is larger. This achieves the greatest possible internal cohesion and external separation. Notice that it does not require preliminary calculation of the distance matrix. Ward's method can be interpreted as a variant of the centroid method, which does require the distance matrix.

How do we choose which method to apply?
In practice, there is no method that can give the best result with every type of data. Experiment the different alternatives and compare them in terms of the chosen criteria. A number of criteria are discussed in the following subsection and more generally in Chapter 5.

Divisive clustering algorithms
The algorithms used for divisive clustering are very similar to those used for tree models (Section 4.5). In general, they are less used in routine applications, as they tend to be more computationally intensive. However, although naïve implementation of divisive methods requires n^2 distance calculations on the first iteration, subsequent divisions are on much smaller cluster sizes. Also, efficient implementations do not compute all pairwise distances but only those that are reasonable candidates for being the closest together.

4.2.2 Evaluation of hierarchical methods

A hierarchical algorithm produces a family of partitions of the n initial statistical units, or better still, a succession of n clusterings of the observations, with the number of groups decreasing from n to 1. To verify that the partitions achieve the primary objective of the cluster analysis – internal cohesion and external separation – the goodness of the partition obtained should be measured at every step of the hierarchical procedure.

A first intuitive criterion for goodness of the clustering is the distance between the joined groups at every step; the process can be stopped when the distance increases abruptly. A criterion used more frequently is based on the decomposition of the total deviance of the p variables, as in Ward's method. The idea is to have a low deviance within the groups (W) and a high deviance between the groups (B). For a partition of g groups here is a synthetic index that expresses this criterion:

$$R^2 = 1 - \frac{W}{T} = \frac{B}{T}.$$

Since $T = W + B$, the index $R^2 \in [0, 1]$; if the value of R^2 approaches 1, it means that the corresponding partition is optimal, since the observations belonging to the same group are very similar (low W) and the groups are well separated (high B). Correspondingly, the goodness of the clustering decreases as R^2 approaches 0.

Note that $R^2 = 0$ when there is only one group and $R^2 = 1$ when there are as many groups as observations. As the number of groups increases, the homogeneity within the groups increases (as each group contains fewer observations), and so does R^2. But this leads to a loss in the parsimony of the clustering. Therefore the maximisation of R^2 cannot be considered the only criterion for defining the optimal number of groups. Ultimately it would lead to a clustering (for which $R^2 = 1$) of n groups, each having one unit.

A common measure to accompany R^2 is the pseudo-F criterion. Let c be a certain level of the procedure, corresponding to a number of groups equal to c, and let n be the number of observations available. The pseudo-F criterion is defined as

$$F_c = \frac{B/(c-1)}{W/(n-c)}.$$

Generally F_c decreases with c since the deviance between groups should decrease and the deviance within groups should increase. If there is an abrupt fall, it means that very different groups are united among them. The advantage of the pseudo-F criterion is that, by analogy with what happens in the context of the normal linear model (Section 4.11), it is possible to show how to build a decision rule that allows us to establish whether to accept the fusion among the groups (null hypothesis) or to stop the procedure, choosing the less parsimonious representation (alternative hypothesis). This decision rule is specified by a confidence interval based on the F distribution, with $(c-1)$ and $(n-c)$ degrees of freedom. But in applying the decision rule, we assume that the observations follow a normal distribution, reducing the advantages of a model-free formulation, such as that adopted here.

An alternative to R^2 is the root mean square standard deviation (RMSSTD). This only considers the part of the deviance in the additional groups formed at each step of the hierarchical clustering. Considering the hth step ($h = 2, \ldots, n - 1$) of the procedure, the RMSSTD is defined as:

$$\text{RMSSTD} = \sqrt{\frac{W_h}{p\,(n_h - 1)}},$$

where W_h is the deviance in the group constituted at step h of the procedure, n_h is its numerosity and p is the number of available variables. A strong increase in RMSSTD from one step to the next shows that the two groups being united are strongly heterogeneous and therefore it would be appropriate to stop the procedure at the earlier step.

Another index that, similar to RMSSTD, measures the 'additional' contribution of the hth step of the procedure is the so-called 'semipartial' R^2 (SPRSQ), given by

$$\text{SPRSQ} = \frac{W_h - W_r - W_s}{T},$$

where h is the new group, obtained at step h as a fusion of groups r and s. T is the total deviance of the observations, while W_h, W_r and W_s are the deviance of the observations in groups h, r and s, respectively. In other words, the SPRSQ measures the increase in the within-group deviance W obtained by joining groups r and s. An abrupt increase in SPSRQ indicates that heterogeneous groups are being united and therefore it is appropriate to stop at the previous step.

We believe that choosing one index from the 'global' indexes R^2 and pseudo-F and one index from the 'local' indexes RMSSTD and SPRSQ allows us to evaluate adequately the degree of homogeneity of the obtained groups in every step of a hierarchical clustering and therefore to choose the best partition.

Table 4.3 gives an example of cluster analysis, obtained with Ward's method, in which the indexes R^2 and SPRSQ are indeed able to give an indication of the

Table 4.3 Output of a cluster analysis.

NCL	Clusters Joined		FREQ	SPRSQ	RSQ
11	CL19	CL24	13	0.0004	0.998
10	CL14	CL18	42	0.0007	0.997
9	CL11	CL13	85	0.0007	0.996
8	CL16	CL15	635	0.0010	0.995
7	CL17	CL26	150	0.0011	0.994
6	CL9	CL27	925	0.0026	0.991
5	CL34	CL12	248	0.0033	0.988
4	CL6	CL10	967	0.0100	0.978
3	CL4	CL5	1215	0.0373	0.941
2	CL7	CL3	1365	0.3089	0.632
1	CL2	CL8	2000	0.6320	0.000

number of partitions to choose. A number of cluster (NCL) equal to 3 is more than satisfactory, as indicated by the row third from last, in which clusters 4 and 5 are united (obtained in correspondence of NCL equal to 4 and 5). In fact, the further step of uniting groups 7 and 3 leads to a relevant reduction in R^2 and to an abrupt increase in SPRSQ. On the other hand, Choosing NCL equal to 4 does not give noticeable improvements in R^2. Note that the cluster joined at NCL = 3 contains 1215 observations (FREQ).

To summarise, there is no unequivocal criterion for evaluating the methods of cluster analysis but a whole range of criteria. Their application should strike a balance between simplicity and information content.

4.2.3 Non-hierarchical methods

The non-hierarchical methods of clustering allow us to obtain one partition of the n observations in g groups ($g < n$), with g defined a priori. Unlike what happens with hierarchical methods, the procedure gives as output only one partition that satisfies determined optimality criteria, such as the attainment of the grouping that allows us to get the maximum internal cohesion for the specified number of groups. For any given value of g, according to which it is intended to classify the n observations, a non-hierarchical algorithm classifies each of the observations only on the basis of the selected criterion, usually stated by means of an objective function. In general, a non-hierarchical clustering can be summarised by the following algorithm:

1. Choose the number of groups g and choose an initial clustering of the n statistical units in that number of groups.
2. Evaluate the 'transfer' of each observation from the initial group to another group. The purpose is to maximise the internal cohesion of the groups. The variation in the objective function determined by the transfer is calculated and, if relevant, the transfer becomes permanent.
3. Repeat step 2 until a stopping rule is satisfied.

Non-hierarchical algorithms are generally much faster than hierarchical ones, because they employ an interactive structure calculation which does not require us to determine the distance matrix. The construction of non-hierarchical algorithms tends to make them more stable with respect to data variability. Furthermore, non-hierarchical algorithms are suitable for large data sets where hierarchical algorithms would be too slow. Nevertheless, there can be many possible ways of dividing n observations into g non-overlapping groups, especially for real data, and it is impossible to obtain and compare all these combinations. This can make it difficult to do a global maximisation of the objective function, and non-hierarchical algorithms may produce constrained solutions, often corresponding to local maxima of the objective function.

In a non-hierarchical clustering we need to begin by defining the number of the groups. This is usually done by conducting the analysis with different values of g (and different algorithm initialisations) and determining the best solution by comparing appropriate indexes for the goodness of the clustering (such as R^2 or the pseudo-F index).

The most commonly used method of non-hierarchical clustering is the k-means method, where k indicates the number of groups established a priori (g in this section). The k-means algorithm performs a clustering of the n starting elements, in g distinct groups (with g previously fixed), according to the following operational flow:

1. **Initialisation.** Having determined the number of groups, g points, called seeds, are defined in the p-dimensional space. The seeds constitute the centroids (measures of position, usually means) of the clusters in the initial partition. There should be sufficient distance between them to improve the properties of convergence of the algorithm. For example, to space the centroids adequately in $\mathbb{R}^p$, one can select g observations (seeds) whose reciprocal distance is greater than a predefined threshold, and greater than the distance between them and the observations. Once the seeds are defined, an initial partition of the observations is constructed, allocating each observation to the group whose centroid is closest.
2. **Transfer evaluation.** The distance of each observation from the centroids of the g groups is calculated. The distance between an observations and the centroid of the group to which it has been assigned has to be a minimum; if it is not a minimum, the observations will be moved to the cluster whose centroid is closest. The centroids of the old group and the new group are then recalculated.
3. **Repetition.** We repeat step 2 until we reach a suitable stabilisation of the groups.

To calculate the distance between the observations and the centroids of the groups, the k-means algorithm employs the Euclidean distance: at the tth iteration, the distance between the ith observation and the centroid of group l (with $i = 1, 2, \ldots, n$ and $l = 1, 2, \ldots, g$) will be equal to

$$d\left(x_i, \bar{x}_l^{(t)}\right) = \sqrt{\sum_{s=1}^{p}\left(x_{is} - \bar{x}_{sl}^{(t)}\right)^2},$$

where $\bar{x}_l^{(t)} = \left[\bar{x}_{1l}^{(t)}, \ldots, \bar{x}_{pl}^{(t)}\right]'$ is the centroid of group l calculated at the tth iteration. This shows that the k-means method searches for the partition of the n observations in g groups (with g fixed in advance) that satisfies a criterion of internal cohesion based on the minimisation of the within-group deviance W, therefore the goodness of the obtained partition can be evaluated by calculating the index R^2 of the pseudo-F statistic. A disadvantage of the k-means method is the possibility of obtaining distorted results when there are outliers in the data. Then the non-anomalous units will tend to be classified into very few groups, but the outliers will tend to be put in very small groups on their own. This can create so-called 'elephant clusters' – clusters too big and containing most of the observations.

4.3 Linear regression

In clustering and, more generally, in descriptive methods, the variables were treated in a symmetric way. We now consider the common situation where we wish to deal with the variables in a non-symmetric way, to derive a predictive model for one (or more) response variables, on the basis of one (or more) of the others. This section focuses on quantitative response variables and the next section focuses on qualitative response variables. Chapter 1 introduced the distinction between descriptive and predictive data mining methods. Linear regression is a predictive data mining method.

We will initially suppose that only two variables are available. Later we will consider the multivariate case.

4.3.1 Bivariate linear regression

In many applications it is interesting to evaluate whether one variable, called the dependent variable or the response, can be caused, explained and therefore predicted as a function of another, called the independent variable, the explanatory variable, the covariate or the feature. We will use Y for the dependent (or response) variable and X for the independent (or explanatory) variable. The simplest statistical model that can describe Y as a function of X is linear regression. The linear regression model specifies a noisy linear relationship between variables Y and X, and for each paired observation (x_i, y_i) this can be expressed by the so-called regression function,

$$y_i = a + bx_i + e_i, \qquad i = 1, 2, \ldots, n,$$

where a is the intercept of the regression function, b is the slope coefficient of the regression function, also called regression coefficient, and e_i is the random error of the regression function, relative to the ith observation.

Note that the regression function has two main parts: the regression line and the error term. The regression line can be constructed empirically, starting with the matrix of available data. The error term describes how well the regression line approximates the observed response variable. From an exploratory point of view, determination of the regression line can be described as a problem of fitting a straight line to the observed dispersion diagram. The regression line is described the linear function:

$$\hat{y}_i = a + bx_i, \qquad i = 1, 2, \ldots, n,$$

where $\hat{y}_i$ denotes the fitted ith value of the dependent variable, calculated on the basis of the ith value of the explanatory variable, x_i. Having defined the regression line, it follows that the error term e_i in the expression for the regression

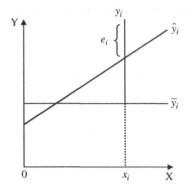

Figure 4.2   Representation of the regression line.

function represents, for each observation y_i, the residual, that is, the difference between the observed response values, y_i, and the corresponding values fitted with the regression line, $\hat{y}_i$:

$$e_i = y_i - \hat{y}_i.$$

Each residual can be interpreted as the part of the corresponding value that is not explained by the linear relationship with the explanatory variable. What we have just described can be represented graphically, as in Figure 4.2. To obtain the analytic expression for the regression line it is sufficient to calculate the parameters a and b on the basis of the available data. The method of least squares is often used for this purpose. It chooses the straight line that minimises the sum of the squares of the errors of the fit (SSE), defined by

$$\text{SSE} = \sum_{i=1}^{n} e_i^2 = \sum_{i=1}^{n} (y_i - \hat{y}_i)^2 = \sum_{i=1}^{n} (y_i - a - bx_i)^2.$$

To find the minimum of SSE we need to take the first partial derivatives of the SSE function with respect to a and b then equate them to zero. Since the sum of the squares of the errors is a quadratic function, if an extremal point exists then it is a minimum. Therefore the parameters of the regression line are found by solving the following system of equations, called normal equations:

$$\frac{\partial \sum (y_i - a - bx_i)^2}{\partial a} = -2 \sum_i (y_i - a - bx_i) = 0,$$

$$\frac{\partial \sum (y_i - a - bx_i)^2}{\partial b} = -2 \sum_i x_i (y_i - a - bx_i) = 0.$$

From the first equation we obtain

$$a = \sum \frac{y_i}{n} - b \sum \frac{x_i}{n} = \mu_Y - b\mu_X.$$

Substituting this into the second equation and simplifying gives

$$b = \frac{\sum x_i y_i / n - \sum y_i \sum x_i / n^2}{\sum x_i^2 / n - \left(\sum x_i / n\right)^2} = \frac{\text{Cov}(X, Y)}{\text{Var}(X)} = r(X, Y)\frac{\sigma_Y}{\sigma_X},$$

where μ_Y and μ_X are the means, σ_Y and σ_X the standard deviations of the variables Y and X, while $r(X,Y)$ indicates the correlation coefficient between X and Y.

Regression is a simple and powerful predictive tool. To use it in real situations, it is only necessary to calculate the parameters of the regression line, according to the previous formulae, on the basis of the available data. Then a value for Y is predicted simply by substituting a value for X into the equation of the regression line. The predictive ability of the regression line is a function of the goodness of fit of the regression line, which is very seldom perfect.

If the variables were both standardised, with zero mean and unit variance, then $a = 0$ and $b = r(X, Y)$. Then $y_i = r(X, Y)x_i$ and the regression line of X, as a function of Y, is simply obtained by inverting the linear relation between Y and X. Even though not generally true, this particular case shows the link between a symmetric analysis of the relationships between variables (described by the linear correlation coefficient) and an asymmetric analysis (described by the regression coefficient b).

Here is a simple regression model for the weekly returns of an investment fund. The period considered goes from 4 October 1994 to 4 October 1999. The objective of the analysis is to study the dependence of the returns on the weekly variations of a stock market index typically used as benchmark (predictor) of the returns themselves, which is named MSCI WORLD.

Figure 4.3 shows the behaviour of a simple regression model for this data set, along with the scatterplot matrix. The intercept parameter a has been set to zero before adapting the model. This was done to obtain a fitted model that would the closest possible to the theoretical financial model known as capital asset pricing model. The slope parameter of the regression line in Figure 4.3 is calculated

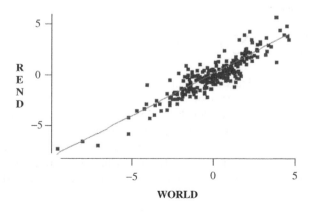

Figure 4.3 Example of a regression line fit.

on the basis of the data, according to the formula previously presented, from which it turns out that $b = 0.8331$. Therefore, the obtained regression line can be analytically described by the equation

$$\text{REND} = 0.8331 \text{ WORLD},$$

where REND is the response variable and WORLD the explanatory variable. The main utility of this model is in prediction: on the basis of the fitted model, we can forecast that if the WORLD index increases by 10% in a week, the fund returns will increase by 8.331%.

4.3.2   Properties of the residuals

We will now look at some important properties of the regression residuals that will permit us to draw some operational conclusions for the diagnostic phase of the model. We will also see an important geometric interpretation of the regression line, an interpretation we can use in the multivariate case. From the first normal equation we have that

$$\sum_{i=1}^{n} e_i = \sum_{i=1}^{n} (y_i - \hat{y}_i) = 0,$$

which shows that the sum of the residuals is null. If in the regression line we set $b = 0$, the arithmetic mean is obtained as a particular linear fit of the dispersion diagram. Such a fit predicts Y with a constant function, ignoring the information provided by X. This property of the regression line coincides, in this particular case, with one of the properties of the arithmetic mean.

From the second normal equation we have that

$$\sum_{i=1}^{n} e_i x_i = \sum_{i=1}^{n} (y_i - \hat{y}_i) x_i = 0.$$

This shows that the residuals are uncorrelated with the explanatory variable. It can also be shown that

$$\sum_{i=1}^{n} e_i \hat{y}_i = \sum e_i (a + b x_i) = a \sum e_i + b \sum e_i x_i = 0,$$

and, therefore, the residuals are uncorrelated also with the fitted Y values.

To investigate the goodness of fit of the regression line, these properties of the residuals suggest that we plot the residuals against the explanatory variable and that we plot the residuals against the fitted values. Both should show a null correlation in mean, and it is interesting to see whether this null correlation is uniform for all the observations considered or whether it arises from compensation of underfit (i.e. the fitted values are smaller than the observed values) and overfit (i.e. the fitted values are greater than the observed values). Compensation of underfit or overfit reduces the validity of the fit. Figure 4.3 shows a uniform distribution of the residuals. There is a slight difference in behaviour between the central part of the regression line, where the variability of the residuals is much larger than in the rest. But this difference does not undermine the excellent fit of the regression line to the data.

The following geometric interpretation is developed for the bivariate case but it can also be extended to the multivariate case. The columns of the data matrix are vectors of dimension n. Therefore they can be thought of as belonging to a linear space. If the variables are quantitative, this space will be the Euclidean space $\mathbb{R}^n$. In the bivariate case under examination in $\mathbb{R}^n$ there will be the vectors $\mathbf{y}$, $\mathbf{x}$ and also $\boldsymbol{\mu} = (1, \ldots, 1)'$, the column vector needed to obtain the arithmetic means of Y and X. In geometric terms, the regression line is a linear combination of two vectors, $\hat{\mathbf{y}} = a\boldsymbol{\mu} + b\mathbf{x}$, determined by two parameters a and b. Therefore it identifies a linear subspace (hyperplane) of $\mathbb{R}^n$ of dimension 2. In general, if we consider k explanatory variables, we obtain a linear subspace of dimension $k + 1$.

To determine a and b we apply the method of least squares. In geometric terms, we determine a vector in $\mathbb{R}^n$ that minimises the Euclidean distance between the observed vector $\mathbf{y}$ in the space $\mathbb{R}^n$ and the estimated vector $\hat{\mathbf{y}}$ belonging to the subspace of dimension $k = 2$ in $\mathbb{R}^n$. The square of this distance is given by

$$d^2(\mathbf{y}, \hat{\mathbf{y}}) = \sum (y_i - \hat{\mathbf{y}}_i)^2.$$

The least squares method minimises the above distance by setting $\hat{\mathbf{y}}$ equal to the projection of the vector $\mathbf{y}$ on the subspace of dimension 2. The properties of the residuals help us to understand what this projection means. The projection $\hat{\mathbf{y}}$ is orthogonal to the vector of the residuals $\mathbf{e}$ (third property). The residuals are orthogonal to $\mathbf{x}$ (second property) and to $\boldsymbol{\mu}$ (first property). We can therefore conclude that the least squares method defines a right-angled triangle, having $\mathbf{y}$ as the hypotenuse and $\hat{\mathbf{y}}$ and $\mathbf{e}$ as the other two sides.

A least squares principle also forms the basis of principal component analysis (Section 3.5); the difference is that in linear regression the distance to be minimised is measured with respect to the response variable only, whereas in principal component analysis it is measured in terms of all variables. We expand on these ideas in the next subsection, on goodness of fit, but first let us see how to interpret the arithmetic mean in geometric terms. The arithmetic mean is definable as the constant quantity, a, that minimises the expression

$$d^2(y, a) = \sum (y_i - a)^2$$

which represents the distance between $\mathbf{y}$ in $\mathbb{R}^n$ and a constant, a, belonging to the subspace of the real numbers of dimension 1 in $\mathbb{R}^n$. Therefore the arithmetic mean is also a solution of the least squares method – it is the projection $\hat{\mathbf{y}}$ of the vector of the observations $\mathbf{y}$ in the subspace $\mathbb{R}$.

4.3.3   Goodness of fit

The regression line represents a linear fit of the dispersion diagram and therefore involves a degree of approximation. We want to measure the accuracy of that approximation. An important judgement criterion is based on a decomposition of the variance of the dependent variable. Recall that the variance is a measure of variability, and variability in statistics means 'information'. By applying Pythagoras' theorem to the right-angled triangle in Section 4.3.2, we obtain

$$\sum (y_i - \bar{y})^2 = \sum (\hat{y}_i - \bar{y})^2 + \sum (y_i - \hat{y}_i)^2.$$

This identity establishes that the total sum of squares (SST), on the left, equals the sum of squares explained by the regression (SSR) plus the sum of squares of the errors (SSE). It can also be written as

$$SST = SSR + SSE .$$

These three quantities are called deviances; if we divide them by the number of observations n, and denote statistical variables using the corresponding capital letters, we obtain

$$\text{Var}(Y) = \text{Var}(\hat{Y}) + \text{Var}(E).$$

We have decomposed the variance of the response variable into two components: the variance 'explained' by the regression line, and the 'residual' variance. This leads to our main index for goodness of fit of the regression line; it is the index of determination R^2, defined by

$$R^2 = \frac{\text{Var}(\hat{Y})}{\text{Var}(Y)} = 1 - \frac{\text{Var}(E)}{\text{Var}(Y)}.$$

The coefficient R^2 is equivalent to the square of the linear correlation coefficient, so it takes values between 0 and 1. It is equal to 0 when the regression line is constant ($Y = \bar{y}$, i.e. $b = 0$); it is equal to 1 when the fit is perfect (the residuals are all null). In general, a high value of R^2 indicates that the dependent variable Y can be well predicted by a linear function of X. The R^2 coefficient of cluster analysis can be derived in exactly the same way by substituting the group means for the fitted line. From the definition of R^2, notice that $\text{Var}(E) = \text{Var}(Y)(1 - R^2)$. This relationship shows how the error in predicting Y reduces from $\text{Var}(Y)$, when the predictor is the mean ($Y = \bar{y}$), to $\text{Var}(E)$, when the predictor is $\hat{y}_i = a + bx_i$. Notice that the linear predictor is at least as good as the mean predictor and its superiority increases with $R^2 = r^2(X, Y)$.

Figure 4.3 has R^2 equal to 0.81. This indicates a good fit of the regression line to the data. For the time being, we cannot establish a threshold value for R^2, above which we can say that the regression is valid, and vice versa. We can do this if we assume a normal linear model, as in Section 4.11.

R^2 is only a summary index. Sometimes it is appropriate to augment it with diagnostic graphical measures, which permit us to understand where the regression line approximates the observed data well and where the approximation is

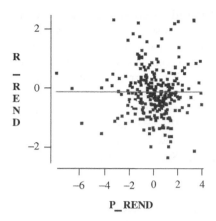

Figure 4.4 Diagnostic plot of a regression model.

poorer. Most of these tools plot the residuals and see what they look like. If the linear regression model is valid, the Y points should be distributed around the fitted line in a random way, without showing obvious trends. It may be a good starting point to examine the plot of the residuals against the fitted values of the response variable. If the plot indicates a difficult fit, look at the plot of the residuals with respect to the explanatory variable and try to see where the explanatory variable is above or below the fit. Figure 4.4 is a diagnostic plot of the residuals (R_REND) against the fitted values (P_REND) for the financial data in Figure 4.3. The diagnostic confirms a good fit of the regression line. Determination of the regression line can be strongly influenced by the presence of anomalous values, or outliers. This is because the calculation of the parameters is fundamentally based on determining mean measures, so it is sensitive to the presence of extreme values. Before fitting a regression model, it is wise to conduct accurate exploratory analysis to identify anomalous observations. Plotting the residuals against the fitted values can support the univariate analysis of Section 3.1 in locating such outliers.

4.3.4 Multiple linear regression

We now consider a more general (and realistic) situation, in which there is more than one explanatory variable. Suppose that all variables contained in the data matrix are explanatory, except for the variable chosen as response variable. Let k be the number of such explanatory variables. The multiple linear regression is defined, for $i = 1, 2, \ldots, n$, by

$$y_i = a + b_1 x_{i1} + b_2 x_{i2} + \ldots + b_k x_{ik} + e_i$$

or, equivalently, in more compact matrix terms,

$$\mathbf{Y} = \mathbf{Xb} + \mathbf{E},$$

where, for all the n observations considered, $\mathbf{Y}$ is a column vector with n rows containing the values of the response variable; $\mathbf{X}$ is a matrix with n rows and $k + 1$ columns containing for each column the values of the explanatory variables for the n observations, plus a column (to refer to the intercept) containing n values equal to 1; $\mathbf{b}$ is a vector with $k + 1$ rows containing all the model parameters to be estimated on the basis of the data (the intercept and the k slope coefficients relative to each explanatory variable); and $\mathbf{E}$ is a column vector of length n containing the error terms. Whereas in the bivariate case the regression model was represented by a line, now it corresponds to a $(k + 1)$-dimensional plane, called the regression plane. Such a plane is defined by the equation

$$\hat{y}_i = a + b_1 x_{i1} + b_2 x_{i2} + \cdots + b_k x_{ik}.$$

To determine the fitted plane it is necessary to estimate the vector of the parameters $(a, b_1, \ldots, b_k)$ on the basis of the available data. Using the least squares optimality criterion, as before, the b parameters will be obtained by minimising the square of the Euclidean distance:

$$d^2(y, \hat{y}) = \sum_{i=1}^{n} (y_i - \hat{y}_i)^2.$$

We can obtain a solution in a similar way to bivariate regression; in matrix terms it is given by $\hat{\mathbf{Y}} = \mathbf{X}\boldsymbol{\beta}$, where

$$\boldsymbol{\beta} = \left(\mathbf{X}'X\right)^{-1} \mathbf{X}'Y.$$

Therefore, the optimal fitted plane results to be defined by

$$\hat{\mathbf{Y}} = \mathbf{X}\left(\mathbf{X}'X\right)^{-1} \mathbf{X}'Y = \mathbf{H}Y.$$

In geometric terms, the previous expression establishes that the optimal plane is obtained as the projection of the observed vector $\mathbf{y} \in \mathbb{R}^n$ on to the $(k + 1)$-dimensional hyperplane. Here the projection operator is the matrix $\mathbf{H}$; in bivariate regression with $a = 0$ the projection operator is b. In fact, for $k = 1$ the two parameters in $\boldsymbol{\beta}$ coincide with parameters a and b in the bivariate case. The properties of the residuals we obtained for the bivariate case can be extended to the multivariate case.

We now apply multiple regression to the investment fund data we have been investigating. We assume a multifactorial model, in conformity with the theory of the arbitrage pricing theory (APT) model. Instead of considering the WORLD index as a unique explanatory variable, we use five indexes relative to specific geographic areas – JAPAN, PACIFIC, EURO, NORDAM, COMIT – as the explanatory variables of the fund return (REND). Table 4.4 summarises the outcome. Notice that the indexes EURO and NORDAM have the strongest effect on the fund return, giving estimated values for the slope regression coefficients that are noticeably greater than the other indexes. For goodness of fit we can still use the variance decomposition identity we obtained for bivariate regression,

$$\text{Var}(Y) = \text{Var}(\hat{Y}) + \text{Var}(E),$$

Table 4.4 Least squares estimates from a multiple regression model.

Variable	Parameter estimate
INTERCEPT	-0.0077
COMIT	-0.0145
JAPAN	0.0716
PACIFIC	0.0814
EURO	0.3530
NORDAM	0.3535

with $\hat{Y}$ now indicating the regression plane fit. This permits us to define the coefficient of multiple determination as a summary index for the plane's goodness of fit:

$$R^2 = \frac{\text{Var}(\hat{Y})}{\text{Var}(Y)} = 1 - \frac{\text{Var}(E)}{\text{Var}(Y)}.$$

The terminology adopted reflects the fact that the plane's goodness of fit depends on the joint effect of the explanatory variables on the response variable. In bivariate regression R^2 is simply the square of the linear correlation coefficient of the response variable with the single explanatory variable; in multivariate regression the relationship is not so straightforward, due to the presence of more than one explanatory variable. An important aim of multivariate regression is to understand not only the absolute contribution of the fitted plane to explaining the variability of Y, as expressed by R^2, but also to determine the partial contribution of each explanatory variable. We now examine in greater detail the variance decomposition identity. It can be demonstrated that

$$\text{Var}(Y) = \sum_{j=1}^{k} b_j \text{Cov}(X_j, Y) + \text{Var}(E).$$

But in general,

$$b_j \neq \frac{\text{Cov}(X_j, Y)}{\text{Var}(X_i)}.$$

If the previous equation were true we would obtain

$$\text{Var}(Y) = \sum_{j=1}^{k} \text{Var}(Y) r^2 (X_j, Y) + \text{Var}(E)$$

and, therefore,

$$\text{Var}(\hat{Y}) = \text{Var}(\hat{Y}_1) + \text{Var}(\hat{Y}_2) + \cdots + \text{Var}(\hat{Y}_k)$$

so that

$$R^2 = \sum_{j=1}^{k} r_{Y,X_j}^2.$$

The variance of Y explained by the fitting plane would be equal to the sum of the variance of Y explained by each of the fitting lines, constructed separately for each of the explanatory variables.

However, this situation occurs only when the explanatory variables are uncorrelated – for example, if the explanatory variables are principal components, obtained using the method in Section 3.5. In general it can be shown that the overall contribution of the fitted plane depends on the single contributions through the recursive relationship

$$R^2 = \sum_{j=1}^{k} r^2_{Y,X_j|X_{i<j}} \left(1 - R^2_{Y,X_1,\dots,X_{j-1}}\right),$$

where $R^2_{Y,X_1,\dots,X_{j-1}}$ denotes the coefficient of multiple correlation between Y and the fitted plane determined by the explanatory variables $X_1, \dots, X_{j-1}$, and $r_{Y,X_j|X_{i<j}}$ denotes the coefficient of partial correlation between Y and X_j, conditional on the 'previous' variables $X_1, \dots, X_{j-1}$. To clarify how his works in practice, consider the case of two explanatory variables ($k = 2$). Fitting first X_1 and then X_2, we get

$$R^2 = r^2_{Y,X_1} + r^2_{Y,X_2|X_1} \left(1 - R^2_{Y,X_1}\right).$$

The term in parentheses takes the amount of variance not explained by the regression (of Y on X_1) and reduces it by a fraction equal to the square of the partial correlation coefficient between itself and the response variable, conditional on the variable X_1 already being present.

To summarise, a single explanatory variable, say X_j, makes an additive contribution to the fitting plane, therefore R^2 increases as the number of variables increases. However, the increase is not necessarily equal to r^2_{Y,X_j}. This occurs only in the uncorrelated case. In general, it can be smaller or greater depending on the degree of correlation of the response variable with those already present, and of the latter with X_j.

When the explanatory variables are correlated, the coefficient of regression estimated for a certain variable can change its sign and magnitude according to the order with which the explanatory variables are inserted in the fitted plane. This can easily be verified with a real application and it emphasises the importance of ordering the explanatory variables. Software packages order the variables according to their predictive capacity, obtained from an exploratory analysis. They might order the variables according to the absolute value of the linear correlation coefficient $r(X, Y)$.

Note the importance of the partial correlation coefficient in explaining an extra variable's contribution to the fitted plane. Consider a fitted plane with k explanatory variables. Suppose we want to add a $(k + 1)$th explanatory variable. The contribution of this variable will be the increase in the variance explained by the plane, from $\text{Var}\left(\hat{Y}_k\right)$ to $\text{Var}\left(\hat{Y}_{k+1}\right)$. This contribution can be measured by the difference

$$\text{Var}\left(\hat{Y}_{k+1}\right) - \text{Var}\left(\hat{Y}_k\right).$$

The square of the coefficient of partial correlation relates this additional contribution to the variance not explained by the fitted plane $\hat{Y}_k$:

$$r^2_{Y, X_{k+1} | X_1, \ldots, X_k} = \frac{\text{Var}\left(\hat{Y}_{k+1}\right) - \text{Var}\left(\hat{Y}_k\right)}{\text{Var}(Y) - \text{Var}(\hat{Y}_k)}.$$

In our financial example the model with the five explanatory variables has a coefficient of multiple determination equal to 0.8191. Among the five considered explanatory variables, the COMIT variable has a coefficient of partial correlation with the response variable, given all the other explanatory variables, equal to about 0.0003. This suggests the possible elimination of the COMIT variable, as it appears substantially irrelevant after having inserted the other explanatory variables. In fact, the coefficient of multiple determination relative to the fit of a model that explains the return as a function of the other four explanatory variables is equal to 0.8189, only slightly inferior to 0.8191.

4.4 Logistic regression

Section 4.3 considered a predictive model for a quantitative response variable; this section considers a predictive model for a qualitative response variable. A qualitative response problem can often be decomposed into binary response problems (e.g. Agresti, 1990). The building block of most qualitative response models is the logistic regression model, one of the most important predictive data mining methods. Let y_i $(i = 1, 2, \ldots, n)$ be the observed values of a binary response variable, which can take only the values 0 or 1. The level 1 usually represents the occurrence of an event of interest, often called a 'success'. A logistic regression model is defined in terms of fitted values to be interpreted as probabilities (Section 4.9) that the event occurs in different subpopulations:

$$\pi_i = P\left(Y_i = 1\right), \qquad i = 1, 2, \ldots, n.$$

More precisely, a logistic regression model specifies that an appropriate function of the fitted probability of the event is a linear function of the observed values of the available explanatory variables. Here is an example:

$$\log\left[\frac{\pi_i}{1 - \pi_i}\right] = a + b_1 x_{i1} + b_2 x_{i2} + \cdots + b_k x_{ik}.$$

The left-hand side defines the logit function of the fitted probability, logit (π_i), as the logarithm of the odds for the event, namely, the natural logarithm of the ratio between the probability of occurrence (success) and the probability of non-occurrence (failure):

$$\text{logit}\left(\pi_i\right) = \log\left[\frac{\pi_i}{1 - \pi_i}\right].$$

Once π_i is calculated, on the basis of the data, a fitted value for each binary observation $\hat{y}_i$ can be obtained, introducing a cut-off threshold value of π_i above which $\hat{y}_i = 1$ and below which $\hat{y}_i = 0$. The resulting fit will seldom be perfect, so there will typically be a fitting error that will have to be kept as low as possible. Unlike linear regression, the observed response values cannot be decomposed additively as the sum of a fitted value and an error term.

The choice of the logit function to describe the link between π_i and the linear combination of the explanatory variables is motivated by the fact that with this choice such probability tends towards 0 and 1 gradually. And these limits are never exceeded, guaranteeing that π_i is a valid probability. A linear regression model would be inappropriate for predicting a binary response variable, for the simple reason that a linear function is unlimited, so the model could predict values for the response variable outside the interval [0,1], which would be meaningless. But other types of link are possible (see Section 4.12).

4.4.1   Interpretation of logistic regression

The logit function implies that the dependence of π_i on the explanatory variables is described by a sigmoid or S-shaped curve. By inverting the expression that defines the logit function, we obtain

$$\pi_i = \frac{\exp(a + b_1 x_{i1} + b_2 x_{i2} + \ldots + b_k x_{ik})}{1 + \exp(a + b_1 x_{i1} + b_2 x_{i2} + \ldots + b_k x_{ik})}.$$

This relationship corresponds to the function known as 'logistic curve', often employed in diffusion problems, including the launch of a new product or the diffusion of a reserved piece of information. These applications usually concern the simple case of only one explanatory variable, corresponding to a bivariate logistic regression model,

$$\pi_i = \frac{e^{a + b_1 x_{i1}}}{1 + e^{a + b_1 x_{i1}}}.$$

Here the value of the success probability varies according to the observed values of the unique explanatory variable. This simplified case is useful to visualise the behaviour of the logistic curve, and to make two more remarks about interpretation. Figure 4.5 shows the graph of the logistic function that links the probability of success π_i to the possible values of the explanatory variable x_i, corresponding to two different signs of the coefficient β. We have assumed the more general setting, in which the explanatory variable is continuous and, therefore, the success probability can be written as $\pi(x)$. In the case of the discrete or qualitative explanatory variables the results will be a particular case of what we will now describe. Notice that the parameter β determines the rate of growth or increase of the curve; the sign of β indicates if the curve increases or decreases and the magnitude of β determines the rate of that increase or decrease:

- When $\beta > 0$, $\pi(x)$ increases as x increases.
- When $\beta < 0$, $\pi(x)$ decreases as x increases.

Furthermore, for $\beta \to 0$ the curve tends to become a horizontal straight line. In particular, when $\beta = 0$, Y is independent of X.

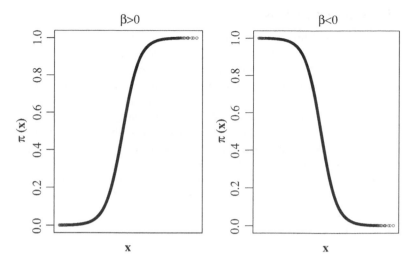

Figure 4.5 The logistic function.

Although the probability of success is a logistic function and therefore not linear in the explanatory variables, the logarithm of the odds is a linear function of the explanatory variables:

$$\log\left(\frac{\pi(x)}{1 - \pi(x)}\right) = \alpha + \beta x.$$

Positive log-odds favour $Y = 1$ whereas negative log-odds favour $Y = 0$. The log-odds expression establishes that the logit increases by β units for a unit increase in x. It could be used during the exploratory phase to evaluate the linearity of the observed logit. A good linear fit of the explanatory variable with respect to the observed logit will encourage us to apply the logistic regression model. The concept of the odds was introduced in Section 3.4. For the logistic regression model, the odds of success can be expressed by

$$\frac{\pi(x)}{1 - \pi(x)} = e^{\alpha + \beta x} = e^{\alpha}(e^{\beta})^{x}.$$

This exponential relationship offers a useful interpretation of the parameter β: a unit increase in x multiplies the odds by a factor e^{β}. In other words, the odds at level $x + 1$ equal the odds at level x multiplied by e^{β}. When $\beta = 0$ we obtain $e^{\beta} = 1$, therefore the odds do not depend on X.

What about the fitting algorithm, the properties of the residuals, and goodness of fit indexes? These concepts can be introduced by interpreting logistic regression as a linear regression model for appropriate transformation of the variables. They are examined as part of the broader field of generalised linear models (Section 4.12), which should make them easier to understand. A real application of the model is also postponed until Section 4.12.

4.4.2   Discriminant analysis

Linear regression and logistic regression models are essentially scoring models: they assign a numerical score to each value to be predicted. These scores can be used to estimate the probability that the response variable assumes a predetermined set of values or levels (e.g. all positive values if the response is continuous or a given level if it is binary). Scores can then be used to classify the observations into disjoint classes. This is particularly useful to classify new observations not already present in the database. This objective is more natural for logistic regression models, where predicted scores can be converted in binary values, thus classifying observations in two classes: those predicted to be 0 and those predicted to be 1. To do this, we need a threshold or cut-off rule. This type of predictive classification rule is studied by the classical theory of discriminant analysis. We consider the simple (and common) case in which each observation is to be classified using a binary response: it is either in class 0 or in class 1. The more general case is similar, but more complex to illustrate.

The choice between the two classes is usually based on a probabilistic criterion: choose the class with the highest probability of occurrence, on the basis of the observed data. This rationale, which is optimal when equal misclassification costs are assumed (Section 4.9), leads to an odds-based rule that allows us to assign an observation to class 1 (rather than class 0) when the odds in favour of class 1 are greater than 1, and vice versa. Logistic regression can be expressed as a linear function of log-odds, therefore a discriminant rule can be expressed in linear terms, by assigning the ith observations to class 1 if

$$a + b_1 x_{i1} + b_2 x_{i2} + \ldots + b_k x_{ik} > 0.$$

With a single predictor variable, the rule simplifies to

$$a + b x_i > 0.$$

This rule is known as the logistic discriminant rule; it can be extended to qualitative response variables with more than two classes.

An alternative to logistic regression is linear discriminant analysis, also known as Fisher's rule. It is based on the assumption that, for each given class of the response variable, the explanatory variables are multivariate normally distributed with a common variance–covariance matrix. Then it is also possible to obtain a rule in linear terms. For a single predictor, the rule assigns observation i to class 1 if

$$\log \frac{n_1}{n_0} - \frac{(\bar{x}_1 - \bar{x}_0)^2}{2s^2} + \frac{x_i(\bar{x}_1 - \bar{x}_0)}{s^2} > 0,$$

where n_1 and n_0 are the number of observations in classes 1 and 0; $\bar{x}_1$ and $\bar{x}_0$ are the observed means of the predictor X in the two classes, 1 and 0; and s^2 is the variance of X for all the observations. Both Fisher's rule and the logistic discriminant rule can be expressed in linear terms, but the logistic rule is simpler to apply and interpret and does not require probability assumptions. Fisher's rule is more explicit than the linear discriminant rule. By assuming a normal distribution, we can add more information to the rule, such as the assessment of its sampling variability.

4.5 Tree models

While linear and logistic regression methods produce a score and then possibly a classification according to a discriminant rule, tree models begin by producing a classification of observations into groups and then obtain a score for each group. Tree models are usually divided into regression trees, when the response variable is continuous, and classification trees, when the response variable is quantitative discrete or qualitative (categorical). However, as most concepts apply equally well to both, here we do not distinguish between them, unless otherwise specified. Tree models can be defined as a recursive procedure, through which a set of n statistical units are progressively divided into groups, according to a division rule that aims to maximise a homogeneity or purity measure of the response variable in each of the obtained groups. At each step of the procedure, a division rule is specified by the choice of an explanatory variable to split and the choice of a splitting rule for the variable, which establishes how to partition the observations.

The main result of a tree model is a final partition of the observations. To achieve this it is necessary to specify stopping criteria for the division process. Suppose that a final partition has been reached, consisting of g groups ($g < n$). Then for any given response variable observation y_i, a regression tree produces a fitted value $\hat{y}_i$ that is equal to the mean response value of the group to which the observation i belongs. Let m be such a group; formally we have that

$$\hat{y}_i = \frac{\sum_{l=1}^{n_m} y_{lm}}{n_m}.$$

For a classification tree, fitted values are given in terms of fitted probabilities of affiliation to a single group. Suppose only two classes are possible (binary classification); the fitted success probability is therefore

$$\pi_i = \frac{\sum_{l=1}^{n_m} y_{lm}}{n_m},$$

where the observations y_{lm} can take the value 0 or 1, and therefore the fitted probability corresponds to the observed proportion of successes in group m. Notice that both $\hat{y}_i$ and π_i are constant for all the observations in the group.

The output of the analysis is usually represented as a tree; it looks very similar to the dendrogram produced by hierarchical clustering (Section 4.2). This implies that the partition performed at a certain level is influenced by the previous choices.

For classification trees, a discriminant rule can be derived at each leaf of the tree. A typical rule is to classify all observations belonging to a terminal node in the class corresponding to the most frequent level (mode). This corresponds to the so-called 'majority rule'. Other 'voting' schemes are also possible but, in the absence of other considerations, this rule is the most reasonable. Therefore each of the leaves points out a clear allocation rule of the observations, which is read by going through the path that connects the initial node to each of them. Every path in a tree model thus represents a classification rule. Compared with discriminant models, tree models produce rules that are less explicit analytically but easier to understand graphically.

Tree models can be considered as non-parametric predictive models, since they do not require assumptions about the probability distribution of the response variable. In fact, this flexibility means that tree models are generally applicable, whatever the nature of the dependent variable and the explanatory variables. But this greater flexibility may have disadvantages, such as a greater demand for computational resources. Furthermore, their sequential nature and their algorithmic complexity can make them dependent on the observed data, and even a small change might alter the structure of the tree. It is difficult to take a tree structure designed for one context and generalise it to other contexts.

Despite their graphical similarities, there are important differences between hierarchical cluster analysis and classification trees. Classification trees are predictive rather than descriptive. Cluster analysis performs an unsupervised classification of the observations on the basis of all available variables, whereas classification trees perform a classification of the observations on the basis of all explanatory variables and supervised by the presence of the response (target) variable. A second important difference concerns the partition rule. In classification trees the segmentation is typically carried out using only one explanatory variable at a time (the maximally predictive explanatory variable), whereas in hierarchical clustering the division (agglomerative) rule between groups is established on the basis of considerations on the distance between them, calculated using all the available variables.

We now describe in more detail the operational choices that have to be made before fitting a tree model to the data. It is appropriate to start with an accurate preliminary exploratory analysis. First, it is necessary to verify that the sample size is sufficiently large. This is because subsequent partitions will have fewer observations, for which the fitted values may have a lot of variance. Second, it is prudent to conduct accurate exploratory analysis on the response variable, especially to identify possible anomalous observations that could severely distort the results of the analysis. Pay particular attention to the shape of the response variable's distribution. For example, if the distribution is strongly asymmetrical, the procedure may lead to isolated groups with few observations from the tail of the distribution. Furthermore, when the dependent variable is qualitative, ideally the number of levels should not be too large. Large numbers of levels should be reduced to improve the stability of the tree and to achieve improved predictive performance.

After the preprocessing stage, choose an appropriate tree model algorithm, paying attention to how it performs. The two main aspects are the division criteria and the methods employed to reduce the dimension of the tree. The most popular algorithm in the statistical community is the CART algorithm (Breiman *et al.*, 1984), which stands for 'classification and regression trees'. Other algorithms include CHAID (Kass, 1980), C4.5 and its later version, C5.0 (Quinlan, 1993).

C4.5 and C5.0 are widely used by computer scientists. The first versions of C4.5 and C5.0 were limited to categorical predictors, but the most recent versions are similar to CART. We now look at two key aspects of the CART algorithm: division criteria and pruning, employed to reduce the complexity of a tree.

4.5.1 Division criteria

The main distinctive element of a tree model is how the division rule is chosen for the units belonging to a group, corresponding to a node of the tree. Choosing a division rule means choosing a predictor from those available, and choosing the best partition of its levels. The choice is generally made using a goodness measure of the corresponding division rule. This allows us to determine, at each stage of the procedure, the rule that maximises the goodness measure. A goodness measure $\Phi(t)$ is a measure of the performance gain in subdividing a (parent) node t according to a segmentation into a number of (child) nodes. Let $t_r, r = 1, \ldots, s$, denote the child groups generated by the segmentation ($s = 2$ for a binary segmentation) and let p_r denote the proportion of observations, among those in t, that are allocated to each child node, with $\sum p_r = 1$. The criterion function is usually expressed as

$$\Phi(s, t) = I(t) - \sum_{r=1}^{s} I(t_r) p_r,$$

where the symbol I denotes an impurity function. High values of the criterion function imply that the chosen partition is a good one. The concept of impurity is used to measure the variability of the response values of the observations. In a regression tree, a node will be pure if it has null variance (all observations are equal) and impure if the variance of the observations is high. More precisely, for a regression tree, the impurity at node m is defined by

$$I_V(m) = \frac{\sum_{l=1}^{n_m} (y_{lm} - \hat{y}_m)^2}{n_m},$$

where $\hat{y}_m$ is the fitted mean value for group m. For regression trees impurity corresponds to the variance; for classification trees alternative measures should be considered. Here we present three choices. The misclassification impurity is given by

$$I_M(m) = \frac{\sum_{l=1}^{n_m} 1(y_{lm}, y_k)}{n_m} = 1 - \pi_k,$$

where y_k is the modal category of the node, with fitted probability π_k, and the function $1(\cdot, \cdot)$ denotes the indicator function, which takes the value 1 if $y_{lm} = y_k$ and 0 otherwise. The Gini impurity is

$$I_G(m) = 1 - \sum_{i=1}^{k(m)} \pi_i^2,$$

where the π_i are the fitted probabilities of the levels present at node m, which are at most $k(m)$. Finally, the entropy impurity is

$$I_E(m) = - \sum_{i=1}^{k(m)} \pi_i \log \pi_i,$$

with π_i as above. Notice that the Gini impurity and entropy impurity correspond to the application of the indexes of heterogeneity (Section 3.1.3) to the observations at node m. Compared to the misclassification impurity, both are more sensitive to changes in the fitted probabilities; they decrease faster than the misclassification rate as the tree grows. Therefore, to obtain a parsimonious tree, choose the misclassification impurity.

Besides giving a useful split criterion, an impurity measure can be used to give an overall assessment of a tree. Let $N(T)$ be the number of leaves (terminal nodes) of a tree T. The total impurity of T is given by

$$I(T) = \sum_{m=1}^{N(T)} I(t_m) p_m$$

where p_m are the observed proportions of observations in the final classification. In particular, the misclassification impurity constitutes a very important assessment of the goodness of fit of a classification tree. Even when the number of leaves coincides with the number of levels of the response variable, it need not be that all the observations classified in the same node actually have the same level of the response variable. The percentage of misclassifications, or the percentage of observations classified with a level different from the observed value, is also called misclassification error or misclassification rate; it is another important overall assessment of a classification tree.

The impurity measure used by CHAID is the distance between the observed and expected frequencies; the expected frequencies are calculated using the hypotheses for homogeneity for the observations in the considered node. This split criterion function is the Pearson X^2 index. If the decrease in X^2 is significant (i.e. the p-value is lower than a prespecified level α) then a node is split; otherwise it remains unsplit and becomes a leaf.

4.5.2 Pruning

In the absence of a stopping criterion, a tree model could grow until each node contains identical observations in terms of values or levels of the dependent variable. This obviously does not constitute a parsimonious segmentation. Therefore it is necessary to stop the growth of the tree at a reasonable dimension. The ideal final tree configuration is both parsimonious and accurate. The first property implies that the tree has a small number of leaves, so that the predictive rule can be easily interpreted. The second property implies a large number of leaves that are maximally pure. The final choice is bound to be a compromise between the two opposing strategies. Some tree algorithms use stopping rules based on thresholds on the number of the leaves, or on the maximum number of steps in the process. Other algorithms introduce probabilistic assumptions on the variables, allowing us to use suitable statistical tests. In the absence of probabilistic assumptions, the growth is stopped when the decrease in impurity is too small. The results of a tree model can be very sensitive to the choice of a stopping rule.

The CART method uses a strategy somewhat different from the stepwise stopping criteria; it is based on the concept of pruning. First the tree is built to its greatest size. This might be the tree with the greatest number of leaves, or the tree in which every node contains only one observation or observations all with the same outcome value or level. Then the tree is 'trimmed' or 'pruned' according to a cost-complexity criterion. Let T be a tree, and let T_0 denote the tree of greatest size. From any tree a subtree can be obtained by collapsing any number of its internal (non-terminal) nodes. The idea of pruning is to find a subtree of T_0 in an optimal way, so as to minimise a loss function. The loss function implemented in the CART algorithm depends on the total impurity of the tree T and the tree complexity:

$$C_\alpha(T) = I(T) + \alpha N(T)$$

where, for a tree T, $I(T)$ is the total impurity function calculated at the leaves, and $N(T)$ is the number of leaves; with α a constant that penalises complexity linearly. In a regression tree the impurity is a variance, so the total impurity can be determined as

$$I(T) = \sum_{m=1}^{N(T)} I_V(m) n_m.$$

We have seen how impurity can be calculated for classification trees. Although any of the three impurity measures can be used, the misclassification impurity is usually chosen in practice. Notice that the minimisation of the loss function leads to a compromise between choosing a complex model (low impurity but high complexity cost) and choosing a simple model (high impurity but low complexity cost). The choice depends on the chosen value of α. For each α it can be shown that there is a unique subtree of T_0 that minimises $C_\alpha(T)$.

A possible criticism of this loss function is that the performance of each tree configuration is evaluated with the same data used for building the classification rules, which can lead to optimistic estimates of the impurity. This is particularly true for large trees, due to the phenomenon we have already seen for regression models: the goodness of fit to the data increases with the complexity, here represented by the number of leaves. An alternative pruning criterion is based on the predictive misclassification errors, according to a technique known as cross-validation (Section 5.4). The idea is to split the available data set, use one part to train the model (i.e. to build a tree configuration), and use the second part to validate the model (i.e. to compare observed and predicted values for the response variable), thereby measuring the impurity in an unbiased fashion. The loss function is thus evaluated by measuring the complexity of the model fitted on the training data set, whose misclassification errors are measured on the validation data set.

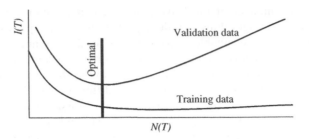

Figure 4.6   Misclassification rates.

To further explain the fundamental difference between training and validation error, Figure 4.6 takes a classification tree and illustrates the typical behaviour of the misclassification errors on the training and validation data sets, as functions of model complexity. $I(T)$ is always decreasing on the training data. $I(T)$ is non-monotone on the validation data; it usually follows the behaviour described in the figure, which allows us to choose the optimal number of leaves as the value of $N(T)$ such that $I(T)$ is minimum. For simplicity, Figure 4.6 takes $\alpha = 0$. When greater values for the complexity penalty are specified, the optimal number of nodes decreases, reflecting aversion to complexity.

The misclassification rate is not the only possible performance measure to use during pruning. Since the costs of misclassification can vary from one class to another, the misclassification impurity could be replaced by a simple cost function that multiplies the misclassification impurity by the costs attached to the consequence of such errors. This is further discussed in Section 5.5.

The CHAID algorithm uses chi-squared testing to produce an implicit stopping criterion based on testing the significance of the homogeneity hypothesis; the hypothesis is rejected for large values of χ^2. If homogeneity is rejected for a certain node, then splitting continues, otherwise the node becomes terminal. Unlike the CART algorithm, CHAID prefers to stop the growth of the tree through a stopping criterion based on the significance of the chi-squared test, rather than through a pruning mechanism.

4.6   Neural networks

Neural networks can be used for many purposes, notably descriptive and predictive data mining. They were originally developed in the field of machine learning to try to imitate the neurophysiology of the human brain through the combination of simple computational elements (neurons) in a highly interconnected system. They have become an important data mining method. However, the neural networks developed since the 1980s have only recently have received the attention from statisticians (e.g. Bishop, Ripley, 1995, 1996). Despite controversies over the real 'intelligence' of neural networks, there is no doubt that they have now become useful statistical models. In particular, they show a notable ability to fit observed data, especially with high-dimensional databases, and data sets characterised by incomplete information, errors or inaccuracies. We will treat neural networks as a methodology for data analysis; we will recall the neurobiological model only to illustrate the fundamental principles.

A neural network is composed of a set of elementary computational units, called neurons, connected together through weighted connections. These units are organised in layers so that every neuron in a layer is exclusively connected to the neurons of the preceding layer and the subsequent layer. Every neuron, also called a node, represents an autonomous computational unit and receives inputs as a series of signals that dictate its activation. Following activation, every neuron produces an output signal. All the input signals reach the neuron simultaneously, so the neuron receives more than one input signal, but it produces only one output signal. Every input signal is associated with a connection weight. The weight determines the relative importance the input signal can have in producing the final impulse transmitted by the neuron. The connections can be exciting, inhibiting or null according to whether the corresponding weights are respectively positive, negative or null. The weights are adaptive coefficients that, by analogy with the biological model, are modified in response to the various signals that travel on the network according to a suitable learning algorithm. A threshold value, called bias, is usually introduced. Bias is similar to an intercept in a regression model.

In more formal terms, a generic neuron j, with a threshold θ_j, receives n input signals $\mathbf{x} = [x_1, x_2, \ldots, x_n]$ from the units to which it is connected in the previous layer. Each signal is attached with an importance weight $\mathbf{w}_j = [w_{1j}, w_{2j}, \ldots, w_{nj}]$.

The same neuron elaborates the input signals, their importance weights, and the threshold value through a combination function. The combination function produces a value called the potential, or net input. An activation function transforms the potential into an output signal. Figure 4.7 schematically represents the activity of a neuron. The combination function is usually linear, therefore the potential is a weighted sum of the input values multiplied by the weights of the respective connections. The sum is compared with the threshold value. The potential of a neuron j is defined by the linear combination

$$P_j = \sum_{i=1}^{n} \left(x_i w_{ij} - \theta_j \right).$$

To simplify this expression, the bias term can be absorbed by adding a further input with constant value, $x_0 = 1$, connected to the neuron j through a weight $w_{0j} = -\theta_j$:

$$P_j = \sum_{i=0}^{n} x_i w_{ij}.$$

Now consider the output signal. The output of the jth neuron, y_j, is obtained by applying the activation function to the potential P_j:

$$y_j = f\left(\mathbf{x}, \mathbf{w}_j\right) = f\left(P_j\right) = f\left(\sum_{i=0}^{n} x_i w_{ij}\right).$$

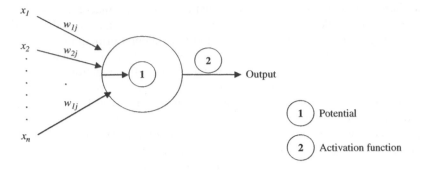

Figure 4.7   Representation of the activity of a neuron in a neural network.

The quantities in bold are vectors.

In the definition of a neural network model, the activation function is typically one of the elements to specify. Three types are commonly employed: linear, stepwise and sigmoidal. A linear activation function is defined by

$$f\left(P_j\right) = \alpha + \beta P_j,$$

where P_j is defined on the set of real numbers, and α and β real constants; $\alpha = 0$ and $\beta = 1$ is a particular case called the identity function, usually employed when the model requires the output of a neuron is exactly equal to its level of activation (potential). Notice the strong similarity between the linear activation function and the expression for a regression line (Section 4.3). In fact, a regression model can be seen as a simple type of neural network.

A stepwise activation function is defined by

$$f\left(P_j\right) = \begin{cases} \alpha, & P_j \geq \theta_j, \\ \beta, & P_j < \theta_j. \end{cases}$$

It can assume only two values depending on whether or not the potential exceeds the threshold θ_j. For $\alpha = 1$, $\beta = 0$ and $\theta_j = 0$ we obtain the so-called sign activation function, which takes the value 0 if the potential is negative and +1 if the potential is positive.

Sigmoidal, or S-shaped, activation functions are probably the most commonly used. They produce only positive output; the domain of the function is the interval [0, 1]. They are widely used because they are non-linear and also because they are easily differentiable and understandable. A sigmoidal activation function is defined by

$$f\left(P_j\right) = \frac{1}{1 + e^{-\alpha P_j}},$$

where α is a positive parameter that regulates the slope of the function.

Another type of activation function, the softmax function, is typically used to normalise the output of different but related nodes. Consider g of such nodes, and let their outputs be v_j, $j = 1, \ldots, g$. The softmax function normalises the v_j so that they sum to 1:

$$\text{softmax}\left(v_j\right) = \frac{\exp\left(v_j\right)}{\sum_{h=1}^{g} \exp\left(v_h\right)}, \qquad j = 1, \ldots, g.$$

The softmax function is used in supervised classification problems, where the response variable can take g alternative levels.

4.6.1 Architecture of a neural network

The neurons of a neural network are organised in layers. These layers can be of three types: of input, of output or hidden. The input layer receives information only from the external environment; each neuron in it usually corresponds to an explanatory variable. The input layer does not perform any calculation; it transmits information to the next level. The output layer produces the final results, which are sent by the network to the outside of the system. Each of its neurons corresponds to a response variable. In a neural network there are generally two or more response variables. Between the output and the input layer there can be one or more intermediate levels, called hidden layers because they are not directly in contact with the external environment. These layers are exclusively for analysis; their function is to take the relationship between the input variables and the output variables and adapt it more closely to the data. In the literature there is no standard convention for calculating the number of layers in a neural network. Some authors count all the layers of neurons and others count the number of layers of weighted neurons. We will use the weighted neurons and count the number of layers that are to be learnt from the data. The 'architecture' of a neural network refers to the network's organisation: the number of layers, the number of units (neurons) belonging to each layer, and the manner in which the units are connected. Network architecture can be represented using a graph, hence people often use the term 'network topology' instead of 'network architecture'. Four main characteristics are used to classify network topology:

- degree of differentiation of the input and output layer;
- number of the layers;
- direction of the flow of computation;
- type of connections.

The simplest topology is called autoassociative; it has a single layer of intra-connected neurons. The input units coincide with the output units; there is no differentiation. We will not consider this type of network, as it is of no statistical interest. Networks with a single layer of weighted neurons are known as single layer perceptrons. They have n input units $(x_1, \ldots, x_n)$ connected to a layer of p output units $(y_1, \ldots, y_p)$, through a system of weights. The weights can be represented in matrix form:

$$
\begin{bmatrix}
w_{11} & \cdots & w_{1j} & \cdots & w_{1p} \\
\vdots & \vdots & \vdots & \vdots & \vdots \\
w_{i1} & \cdots & w_{ij} & \cdots & w_{ip} \\
\vdots & \vdots & \vdots & \vdots & \vdots \\
w_{n1} & \cdots & w_{nj} & \cdots & w_{np}
\end{bmatrix},
$$

for $i = 1, \ldots, n$ and $j = 1, \ldots, p$; w_{ij} is the weight of the connection between the ith neuron of the input layer and the jth neuron of the output layer.

Neural networks with more than one layer of weighted neurons, which contain one or more hidden layers, are called multilayer perceptrons, and we will concentrate on these. A two-layer network has one hidden layer; there are n neurons in the input layer, h in the hidden layer and p in the output layer. Weights w_{ik} ($i = 1, \ldots, n; k = 1, \ldots, h$) connect the input layer nodes with the hidden layer nodes; weights z_{kj} ($k = 1, \ldots, h; j = 1, \ldots, p$) connect the hidden layer nodes with the output layer nodes. The neurons of the hidden layer receive information from the input layer, weighted by the weights w_{ik}, and produce outputs $h_k = f(\mathbf{x}, \mathbf{w}_k)$, where f is the activation function of the units in the hidden layer. The neurons of the output layer receive the outputs from the hidden layer, weighted by the weights z_{kj}, and produce the final network outputs $y_j = g(\mathbf{h}, \mathbf{z}_j)$. The output of neuron j in the output layer is

$$
y_j = g\left(\sum_k h_k z_{kj}\right) = g\left(\sum_k z_{kj} f\left(\sum_i x_i w_{ik}\right)\right).
$$

This equation shows that the output values of a neural network are determined recursively and typically in a non-linear way.

Different information flows lead to different types of network. In feedforward networks the information moves in only one direction, from one layer to the next, and there are no return cycles. In feedback networks it is possible for information to return to previous layers. If each unit of a layer is connected to all the units of the next layer, the network is described as totally interconnected; if each unit is connected to every unit of every layer, the network is described as totally connected.

Networks can also be classified into three types according to their connection weightings: networks with fixed weights, supervised networks and unsupervised networks. We shall not consider networks with fixed weights as they cannot learn from the data and they cannot constitute a statistical model. Supervised networks use a supervising variable, a concept introduced in Section 4.5. With a supervised network, there can be information about the value of a response variable corresponding to the values of the explanatory variables; this information can be used to learn the weights of the neural network model. The response variable behaves as a supervisor for the problem. When this information is not available, the learning of the weights is exclusively based on the explanatory variables and there is no supervisor. Here is the same idea expressed more formally:

- **Supervised learning.** Assume that each observation is described by a pair of vectors $(\mathbf{x}_i, \mathbf{t}_i)$ representing the explanatory and response variables, respectively. Let $D = \{(\mathbf{x}_1, \mathbf{t}_1), \ldots, (\mathbf{x}_n, \mathbf{t}_n)\}$ represent the set of all available observations. The problem is to determine a neural network $\mathbf{y}_i = f(\mathbf{x}_i)$, $i = 1, \ldots, n$, such that the sum of the distances $d(\mathbf{y}_i, \mathbf{t}_i)$ is minimum. Notice the analogy with linear regression models.
- **Unsupervised learning.** Each observation is described by only one vector, with all available variables, $D = \{(\mathbf{x}_1), \ldots, (\mathbf{x}_n)\}$. The problem is the partitioning of the set D into subsets such that the vectors $\mathbf{x}_i$ belonging to the same subset are 'close' in comparison to a fixed measure of distance. This is basically a classification problem.

We now examine the multilayer perceptron, an example of a supervised network, and the Kohonen network, an example of an unsupervised network.

4.6.2 The multilayer perceptron

The multilayer perceptron is the most commonly used architecture for predictive data mining. It is a feedforward network, with possibly several hidden layers, one input layer and one output layer, totally interconnected. It can be considered as a highly non-linear generalisation of the linear regression model when the output variables are quantitative, or of the logistic regression model when the output variables are qualitative.

Preliminary analysis
Multilayer perceptrons, and the neural networks in general, are often used inefficiently on real data because no preliminary considerations are applied. Although neural networks are powerful computation tools for data analysis, they also require exploratory analysis (Chapter 3).

Coding of the variables
The variables used in a neural networks can be classified according by type (qualitative or quantitative) and by their role in the network (input or output). Input and output in neural networks correspond to explanatory and response variables in statistical methods. In a neural network, quantitative variables are represented by one neuron. Qualitative variables, both explanatory and response, are represented in a binary way using several neurons for every variable; the number of neurons equals the number of levels of the variable. In practice the number of neurons to represent a variable need not be exactly equal to the number of its levels. It is advisable to eliminate one level, and therefore one neuron, since the value of that neuron will be completely determined by the others.

Transformation of the variables

Once the variables are coded, a preliminary descriptive analysis may indicate the need for some kind of transformation, perhaps to standardise the input variables to weight them in a proper way. Standardisation of the response variable is not strictly necessary. If a network has been trained with transformed input or output, when it is used for prediction, the outputs must be mapped on to the original scale.

Reduction of the dimensionality of the input variables

One of the most important forms of preprocessing is reduction of the dimensionality of the input variables. The simplest approach is to eliminate a subset of the original inputs. Other approaches create linear or non-linear combinations of the original variables to represent the input for the network. Principal component methods can be usefully employed here (Section 3.5).

Choice of the architecture

The architecture of a neural network can have a fundamental impact on real data. Nowadays, many neural networks optimise their architecture as part of the learning process. Network architectures are rarely compared using the classical methods described later in this chapter; this is because a neural network does not need an underlying probabilistic model and seldom has one. Even when there is an underlying probabilistic model, it is often very difficult to draw the distribution of the statistical test functions. Instead it is possible to make comparison based on the predictive performance of the alternative structures; an example is the cross-validation method (Section 5.4).

Learning of the weights

Learning the weights in multilayer perceptrons appears to introduce no particular problems. Having specified an architecture for the network, the weights are estimated on the basis of the data, as if they were parameters of a (complex) regression model. But in practice there are at least two aspects to consider:

- The error function between the observed values and the fitted values could be a classical distance function, such as the Euclidean distance or the misclassification error, or it could depend in a probabilistic way on the conditional distribution of the output variable with respect to the inputs.
- The optimisation algorithm needs to be a computationally efficient method to obtain estimates of the weights by minimising the error function.

The error functions usually employed for multilayer perceptrons are based on the maximum likelihood principle (Section 4.9). For a given training data set $D = \{(\mathbf{x}_1, \mathbf{t}_1), \ldots, (\mathbf{x}_n, \mathbf{t}_n)\}$, this requires us to minimise the entropy error function

$$E(\mathbf{w}) = -\sum_{i=1}^{n} \log p\,(\mathbf{t}_i\,|\mathbf{x}_i\,;\,\mathbf{w}),$$

where $p(\mathbf{t}_i | \mathbf{x}_i ; \mathbf{w})$ is the distribution of the response variable, conditional on the values of the input values and weighting function. For more details, see Bishop (1995). We now look at the form of the error function for the two principal applications of the multilayer perceptron: predicting a continuous response (predictive regression) and a qualitative response (predictive classification).

Error functions for predictive regression

Every component $t_{i,k}$ of the response vector $\mathbf{t}_i$ is assumed to be the sum of a deterministic component and an error term, similar to linear regression:

$$t_{i,k} = y_{i,k} + \varepsilon_{i,k}, \qquad k = 1, \ldots, q,$$

where $y_{i,k} = \mathbf{y}(\mathbf{x}_i, \mathbf{w})$ is the kth component of the output vector $\mathbf{y}_i$. To obtain more information from a neural network for this problem it can be assumed that the error terms are normally distributed, similar to the normal linear model (Section 4.11).

Since the objective of statistical learning is to minimise the error function in terms of the weights, we can omit everything that does not depend on the weights. We obtain

$$E(\mathbf{w}) = \sum_{i=1}^{n} \sum_{k=1}^{q} (t_{i,k} - y_{i,k})^2.$$

This expression can be minimised using a least squares procedure (Section 4.3). In fact, linear regression can be seen as a neural network model without hidden layers and with a linear activation function.

Error functions for predictive classification

Multilayer perceptrons can also be employed for solving classification problems. In this type of application, a neural network is employed to estimate the probabilities of affiliation of every observation to the various groups. There is usually an output unit for each possible class, and the activation function for each output unit represents the conditional probability $P(C_k | \mathbf{x})$, where C_k is the kth class and $\mathbf{x}$ is the input vector. The output value $y_{i,k}$ represents the fitted probability that the observation i belongs to the kth group C_k. To minimise the error function with respect to the weights, we need to minimise

$$E(\mathbf{w}) = - \sum_{i=1}^{n} \sum_{k=1}^{q} \left[t_{i,k} \log y_{i,k} + (1 - t_{i,k}) \log (1 - y_{i,k}) \right],$$

which represents a distance based on the entropy index of heterogeneity (Section 3.1). Notice that a particular case of the preceding expression can be obtained for the logistic regression model.

In fact, logistic regression can be seen as a neural network model without hidden nodes and with a logistic activation function and softmax output function. In contrast to logistic regression, which produces a linear discriminant rule, a multilayer perceptron provides a non-linear discriminant rule which is not amenable to simple analytical description.

Choice of optimisation algorithm

In general, the error function $E(\mathbf{w})$ of a neural network is a function highly non-linear in the weights, so there may be many minima that satisfy the condition $\nabla E = 0$. Consequently, it may not be possible, in general, to find a globally optimal solution, $\mathbf{w}^*$. Therefore, we must resort to iterative algorithms. We guess an initial estimate $\mathbf{w}^{(0)}$, then produce a sequence of points $\mathbf{w}^{(s)}$, $s = 1, 2, \ldots$, that converge to a certain value $\hat{\mathbf{w}}$. Here are the steps in more detail:

1. Choose a direction $d^{(s)}$ for the search.;
2. Choose a width (or momentum) $\alpha^{(s)}$ and set $\mathbf{w}^{(s+1)} = \mathbf{w}^{(s)} + \alpha^{(s)} d^{(s)}$.
3. If a certain criterion of convergence is satisfied then set $\hat{\mathbf{w}} = \mathbf{w}^{(s+1)}$, otherwise set $s = s + 1$ and return to step 1.

Iterative methods guarantee convergence towards minimum points for which $\nabla E = 0$. Different algorithms have different ways of changing the vector of weights $\Delta \mathbf{w}^{(s)} = \alpha^{(s)} d^{(s)}$. A potential problem for most of them is getting stuck in a local minimum; the choice of the initial weights determines the minimum to which the algorithm will converge. It is extremely important to choose the weights carefully to obtain a valid fit and a good convergence rate. The momentum parameter also needs to be chosen carefully. If it is too small, the algorithm will converge too slowly; if it is too large, the algorithm will oscillate in an unstable way and may never converge.

One last choice for the analyst is when to interrupt the learning algorithm. Here are some possibilities: stop after a defined number of iterations; stop after a defined amount of computer time (CPU usage); stop when the error function falls below a preset value; stop when the difference between two consecutive values of the error function is less than a preset value; stop when the error of classification or forecast, measured on an appropriate validation data set, starts to grow (early stopping), similar to tree pruning (Section 4.5). For more details, see Bishop (1995). It is not possible to establish in general which is the best algorithm; performance varies from problem to problem.

Generalisation and prediction

The objective of training a neural network with data, that is, to determine its weights on the basis of the available data set, is not to find an exact representation of the training data, but to build a model that can be generalised or that allows us to obtain valid classifications and predictions when fed with new data. Similar to tree models, the performance of a supervised neural network can be evaluated with reference to a training data set or a validation data set. If the network is very complex, and the training is carried out for a large number of iterations, the network can perfectly classify or predict the data in the training set. This could be desirable when the training sample represents a 'perfect' image of the population from which it has been drawn, but it is counterproductive in real applications since it implies reduced predictive capacities on a new set of data. This phenomenon is known as overfitting. To illustrate the problem, consider only two observations for an input variable and an output variable. A straight line fits the data perfectly, but poorly predicts a third observation, especially if it is radically different from the previous two. A simpler model, the arithmetic mean of the two output observations, will fit the two points worse but may be a reasonable predictor of the third point.

To limit the overfitting problem, it is important to control the degree of complexity of the model. A model with few parameters will involve a modest generalisation. A model which is too complex may even adapt to noise in the data set, perhaps caused by measurement errors or anomalous observations; this will lead to inaccurate generalisations. There are two main approaches to controlling the complexity of a neural network. Regularisation is the addition of a penalty term to the error function. Early stopping is the introduction of stopping criteria in the iterative learning procedure.

In regularisation, overfitting is tackled directly when the weights are estimated. More precisely, the weights are trained by minimising an error function of the form

$$\tilde{E}(\mathbf{w}) = E(\mathbf{w}) + v\Omega,$$

where E is an error function, Ω describes the complexity of the network and v is a parameter that penalises for complexity. Notice again the analogies with pruning in tree models (Section 4.5). A complex network that produces a good fit to the training data will show a minimum value of E, whereas a very simple function will have low value of Ω. Therefore, what will be obtained at the end of the training procedure will be a compromise between a simple model and a good fit to the data. A useful regularisation function is based on weight decay, which involves taking Ω equal to the sum of the squares of the weights (including the bias) of the neural network:

$$\Omega = \frac{1}{2} \sum_i w_i^2.$$

As an alternative to regularisation, early stopping uses the fact that the error function usually shows an initial reduction followed by an increase; the increase starts when the network begins to have problems with overfitting. Training can be stopped when the lowest prediction error is observed.

Optimality properties of multilayer perceptrons

Multilayer perceptrons have optimal properties. Researchers have shown that, given a sufficiently large number of nodes in the hidden layer, a simple neural network structure (with two layers of weights, sigmoidal activation function for the hidden nodes and identity activation function for the output nodes) is able to approximate any functional form with arbitrary accuracy. This is known as the principle of universal approximation – the rate of convergence does not depend on the dimension of the problem. If a network with only one hidden layer can approximate any functional form with arbitrary accuracy, why use any other network topology? One reason is that extra hidden layers may produce a more efficient approximation that achieves the same level of accuracy using fewer neurons and fewer weights.

Application

Here is a simple example to illustrate the application of neural networks. The data set is a sample of 51 enterprises in the European software industry, for which a series of binary variables have been measured. Here are some of them: N (degree of incremental innovations: low/high); I (degree of radical innovations: low/high); S (relationships of the enterprise with the software suppliers: low/high); A (applicable knowledge of the employees of the enterprise: low/high); M (scientific knowledge of the employees of the enterprise: low/high); H (relationships of the enterprise with the hardware suppliers: low/high). The variable Y (revenues) is a continuous variable.

The objective of the analysis is to classify the 51 enterprises in the two groups of the variable N, according to the values of the six remaining (explanatory) variables, so to build a predictive model for the degree of incremental innovations. Since we have only one response variable and six explanatory variables, the network architecture will have one output variable and six input variables. It remains to be seen how many neurons should be allocated to the hidden units. Suppose that, for parsimony, there is only one hidden node in a unique hidden layer. Finally, given the nature of the problem, a logistic activation function is chosen for the hidden layer node and an identity activation function for the output nodes. The following formula specifies the non-linear relationship for the model:

$$\text{logit}(\pi_N) = w_{08} + w_{18}m + w_{28}a + w_{38}h + w_{48}i + w_{58}s + w_{68}y + w_{78}\phi$$

$$(w_{07} + w_{17}m + w_{27}a + w_{37}h + w_{47}i + w_{57}s + w_{67}y),$$

where the left-hand side is the logit function for $N = 1$ and ϕ is the inverse logistic function. Notice that a logistic regression model differs from this one in not having the term $w_{78}\phi(\cdot)$. The process of learning the weights converges and produces the following 15 final weights:

$$\hat{\omega}_{07} = 32.76, \hat{\omega}_{17} = 9.25, \hat{\omega}_{27} = 14.72, \hat{\omega}_{37} = 3.63, \hat{\omega}_{47} = -10.65,$$

$$\hat{\omega}_{57} = 10.39, \hat{\omega}_{67} = -22.34, \hat{\omega}_{08} = 0.06, \hat{\omega}_{78} = 10.89, \hat{\omega}_{18} = -1.44,$$

$$\hat{\omega}_{28} = -0.82, \hat{\omega}_{38} = -2.18, \hat{\omega}_{48} = -0.70, \hat{\omega}_{58} = -8.34, \hat{\omega}_{68} = 0.43.$$

As a simple measure of its performance, consider the number of misclassified observations. Given the limited number of observations, at first the model was initially trained and validated on the whole data set. Then, the 51 observations were randomly divided into 39 observations for training and 12 observations for validation by considering the misclassifications of these 12 observations. Adopting the threshold rule that $N = 1$ if the estimated value of π_N is greater than 0.5, the number of misclassifications is 9 on 51 training cases and 8 on 12 validation cases. The proposed model is too adaptive. It performs well on training error but very poorly on classification ability. Application of a simpler logistic regression model led to 15 out of 51 and 3 out of 12 errors. So compared with the neural network model, the logistic regression is not as adaptive but more predictive.

4.6.3 Kohonen networks

Self-organising maps (SOMs), or Kohonen networks, can be employed in a descriptive data mining context, where the objective is to cluster the observations into homogeneous groups. In these models the parameters are constituted by the weights of the net (the thresholds are not present) and learning occurs in the absence of an output variable acting as supervisor. A model of this type is generally specified by a layer of input neurons and a layer of output neurons. For a given set of n observations, the n input nodes are represented by p-dimensional vectors (containing qualitative and/or quantitative variables), each of which represents one multivariate observation, whereas the output nodes are described by discrete values, each of which corresponds to a group (cluster) of the observations. The number of groups is typically unknown a priori.

The objective of Kohonen maps is to map every p-dimensional input observation to an output space represented by a spatial grid of output neurons. Adjacent output nodes will be more similar than distant output nodes. The learning technique for the weights in a Kohonen map is based on competition among the output neurons for assignment of the input vectors. For every assigned input node, a neuron of winning output is selected on the basis of the distance function in the input space.

Kohonen networks can be considered as a non-hierarchical method of cluster analysis. As non-hierarchical methods of clustering, they assign an input vector to the nearest cluster, on the basis of a predetermined distance function, but they try to preserve a degree of dependence among the clusters by introducing a distance between them. Consequently, each output neuron has its own neighbourhood, expressed in terms of a distance matrix. The output neurons are characterised by a distance function between them, described using the configuration of the nodes in a one- or two-dimensional space. Figure 4.8 shows a two-dimensional grid of output neurons. In such a 7×7 map, each neuron is described by a square and the number on each square is the distance from the central neuron. Consider the simplest algorithm, in which the topological structure of the output nodes is constant during the learning process. Here are the basic steps:

1. **Initialisation.** Having fixed the dimensions of the output grid, the weights that connect the input neurons to the output neurons are randomly initialised. Let r be the number of iterations of the algorithm, and set $r = 0$.

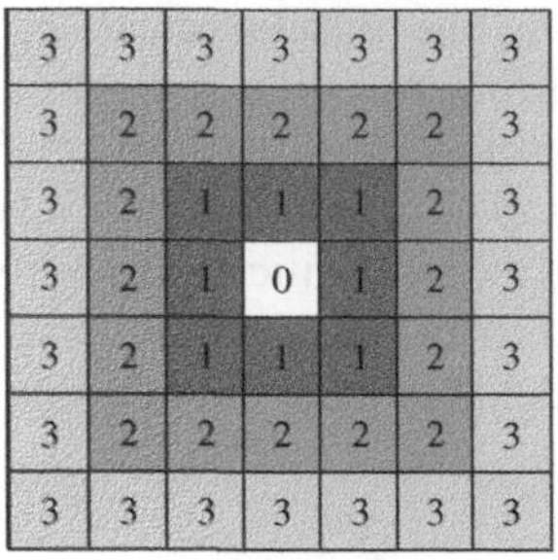

Figure 4.8   Example of output grid in a Kohonen network.

2. **Selection of the winner.** For each input neuron x_j, select the winning output neuron i^* that minimises the Euclidean distance $\|x_j - w_i^r\|$ between the p-dimensional vector of input x_j and the p-dimensional weight vector w_i that connects the jth input neuron of to the ith output neuron.

3. **Updating of the weights.** Let $N(i^*)$ be a neighbourhood of the winning output neuron i^*, implicitly specified by the distance function among the output neurons. For every output neuron $i \in \{N(i^*), i^*\}$, the weights are updated according to the rule $w_i^{r+1} = w_i^r + \eta(x_j - w_i^r)$, where η is called the rate of learning and is specified in advance. The rule updates only the neighbours of the winning output neuron.

4. **Normalisation of the weights.** After the updating, the weights are normalised so that they are consistent with the input measurement scales.

5. **Looping through.** The preceding steps are repeated, and the number of iterations set to $r = r + 1$, until an appropriately stopping criterion is reached or a maximum number of iterations is exceeded.

This algorithm can be modified in at least two important ways. One way is to introduce a varying neighbourhood. After selecting the winning output neuron, its neighbourhood is recomputed along with the relevant weights. Another way is to introduce algorithms based on sensitivity to history. Then the learning algorithm, hence the cluster allocation, can be made to depend on the frequency of past allocations. This allows us to avoid phenomena that typically occur with non-hierarchical clustering, such as obtaining one enormous cluster compared to the others.

SOMs are an important methodology for descriptive data mining and they represent a valid alternative to clustering methods. They are closely related to non-hierarchical clustering algorithms, such as the k-means method. The fundamental difference between the two methodologies is that SOM algorithms introduce a topological dependence between clusters. This can be extremely important when it is fundamental to preserve the topological order among the input vectors and the clusters. This is what happens in image analysis, where it is necessary to preserve a notion of spatial correlation between the pixels of the image. Clustering methods may overcentralise, since the mutual independence of the different groups leads to only one centroid being modified, leaving the centroids of the other clusters unchanged; this means that one group gets bigger and bigger while the other groups remain relatively empty. But if the neighbourhood of every neuron is so small as to contain only one output neuron, the SOMs will behave analogously to the k-means algorithm.

4.7   Nearest-neighbour models

Nearest-neighbour methods are a flexible class of predictive data mining methods based on a combination of local models. This does not mean that they are local in the sense of Section 4.8; they are still applied to the whole data set, but the statistical analysis is split into separate local analyses. The basic idea is rather simple and builds on the theory we have presented in previous sections. The available variables are divided into the explanatory variables ($\mathbf{x}$) and the target variable (y). A sample of observations in the form ($\mathbf{x}$, y) is collected to form a training data set. For this training data set, a distance function is introduced between the $\mathbf{x}$ values of the observations. This can be used to define, for each observation, a neighbourhood formed by the observations that are the closest to it, in terms of the distance between the $\mathbf{x}$ values. For a continuous response variable, the nearest-neighbour fitted value for each observation's response value y_i is defined by

$$\hat{y}_i = \frac{1}{k} \sum_{x_j \in N(x_i)} y_j.$$

This is the arithmetic mean of all response values, whose corresponding $\mathbf{x}$ values are contained in the neighbourhood of $\mathbf{x}_i$, $N(x_i)$. Furthermore, k is a fixed constant, that specifies the number of elements to be included in each neighbourhood. The model can be easily applied to predict a future value of y, say y_0, when the values of the explanatory variables, say $\mathbf{x}_0$, are known. It is required to identify, in the training data set, the k values of y belonging to the neighbourhood of the unknown y_0. This is done by taking the k explanatory variable observations in the training data set, closest to $\mathbf{x}_0$. The arithmetic mean of these y values is the prediction of y_0. In contrast with linear regression, the nearest-neighbour fit is simpler, as it is an arithmetic mean. However, it is not calculated over all observation points, but on a local neighbourhood. This implies that the nearest-neighbour model fits the data more closely; on the other hand, this may lead to overfitting and difficulty with generalisation.

Nearest-neighbour methods can also be used for predictive classification. To classify an observation y, its neighbourhood is determined as before and the fitted probabilities of each category are calculated as relative frequencies in the neighbourhood. The class with the highest fitted probability is finally chosen. Like tree models, nearest-neighbour models do not require a probability distribution. But whereas classification trees partition the data into exclusive classes, providing explicit predictive rules in terms of tree paths, the fitted values in nearest-neighbour models are based on overlapping sets of observations, not on explicit rules. These methods are also known as memory-based models, as they require no model to be fitted, or function to be estimated. Instead they require all observations to be maintained in memory, and when a prediction is required, they recall items from memory and calculate what is required.

Two crucial choices in nearest neighbour-methods are the distance function and the cardinality k of the neighbourhood. Distance functions are discussed in Section 4.1. The cardinality k represents the complexity of the nearest-neighbour model; the higher the value of k, the less adaptive the model. Indeed, the model is often called the k-nearest-neighbour model to emphasise the importance of k. In the limit, when k is equal to the number of observations, the nearest-neighbour fitted values coincide with the sample mean. As we have seen for other models in this chapter (e.g. Sections 4.5 and 4.6), k can be chosen to balance goodness of fit with simplicity.

Possible disadvantages of these models are that computationally they are highly intensive, especially when the data set contains many explanatory variables. In this case the neighbourhood may be formed by distant points, therefore taking their mean may not be a sensible idea. Among other possible data mining applications, they are used for detecting frauds involving telephone calls, credit cards, etc. (Cortes and Pregibon, 2001). Impostors are discovered by identifying the characteristics, or footprints, of previous instances of fraud and formulating a rule to predict future occurrences.

4.8   Local models

So far we have looked at global models, but local models are also very important. They look at selected parts of the data set (subsets of variables or subsets of observations), rather than being applied to the whole data set. Hand *et al.* (2001) use the concept of 'pattern' rather than the concept of 'model'. Relevant examples are association rules, developed in market basket analysis and web clickstream analysis, and retrieval-by-content methods, developed for text mining. Another important example is searching for outliers, introduced in Chapter 3 and revisited several times in this book.

4.8.1   Association rules

Association rules were developed in the field of computer science and are often used in important applications such as market basket analysis, to measure the associations between products purchased by a particular consumer, and web click-stream analysis, to measure the associations between pages viewed sequentially by a website visitor. In general, the objective is to underline groups of items that typically occur together in a set of transactions. The data on which association rules are applied are usually in the form of a database of transactions. For each transaction (a row in the database) the database contains the list of items that occur. Note that each individual may appear more than once in the data set. In market basket analysis a transaction means a single visit to a supermarket, for which the list of purchases is recorded; in web clickstream analysis a transaction means a web session, for which the list of all visited web pages is recorded.

Rows typically have a different number of items, and this is a remarkable difference with respect to data matrices. Alternatively, the database can be converted into a binary data matrix, with transactions as rows and items as columns. Let $X_1, \ldots, X_p$ be a collection of random variables. In general, a pattern for such variables identifies a subset of all possible observations over them. A useful way to describe a pattern is through a collection of primitive patterns and a set of logical connectives that can act on them. For two variables, Age and Income, a pattern could be $\alpha = (\text{Age} < 30 \wedge \text{Income} > 100)$, where $\wedge$ is the logical 'AND' (intersection) operator. Another pattern could be $\beta = (\text{Gender} = \text{male} \vee \text{Education} = \text{high})$, where $\vee$ is the logical 'OR' (union) operator. The primitive patterns in the first expression are Age <30 and Income ⟩100; the primitive patterns in the second expression are Gender = male and Education = high. A rule is a logical statement between two patterns, say α and β, written as $\alpha \rightarrow \beta$. This means that α and β occur together; in other words, if α occurs, then β also occurs. It is an expression of the type 'if condition, then result'.

Association rules consider rules between special types of pattern, called item sets. In an item set, each variable is binary: it takes value 1 if a specific condition is true, otherwise it takes value 0. Let $A_1, \ldots, A_p$ denote a collection of such binary variables, and $j_1, \ldots, j_k$ a subset of them. An item set is then defined by a pattern of the type $A = (A_{j1} = 1 \wedge \ldots \wedge A_{jk} = 1)$. Thus, in an item set, primitive patterns always indicate that a particular variable is true, and the logical connectives are only conjunctions (AND operators). An association rule is a statement between two item sets that can be written in the form $A \rightarrow B$, where both A and B are item sets. For simplicity, the right-hand-side item set is usually formed of a single primitive item, and we will do the same. Therefore an association rule will have the form $(A_{j1} = 1 \wedge \ldots \wedge A_{jk} = 1) \rightarrow A_{jk+1} = 1$, where we have now considered a subset containing $k + 1$ of the original p variables. More briefly, such an association rule is usually written as $(A_{j1} = 1 \wedge \ldots \wedge A_{jk}) \rightarrow A_{jk+1}$.

The order of an association rule usually refers to the total number of items considered, here, $k + 1$. Suppose a supermarket has a total of 100 000 available products. Each of them can correspond to a binary random variable, depending on whether or not the product is bought in each transaction. A simple association rule of order 3 would be (Milk $\wedge$ Tea) $\rightarrow$ Biscuits. We shall simply write $A \rightarrow B$ to indicate an association rule of the described type. A is the antecedent (or body of the rule) and B is the consequent (or head of the rule). Chapters 7 and 8 consider specific applications and use real variables.

Each association rule describes a particular local pattern that selects a restricted set of binary variables. In market basket analysis and web clickstream analysis, rules are relationships between variables that are binary by nature. This need not always be the case; continuous rules are also possible. Then the elements of the rules would be intervals of the real line, conventionally assigned a value of TRUE = 1. A rule of this kind is $X > 0 \rightarrow Y > 100$. Here we shall be mainly concerned with binary variables. The strength of an association rule is commonly measured using support, confidence and lift, also known as measures of a rule's 'statistical interestingness' (Hand *et al.*, 2001).

The main problem in association rule modelling is to find, from the available database, a subset of association rules that are interesting. Interestingness can be measured by various means, including subject-matter criteria and objective-driven criteria. Here we consider statistical interestingness, which is related to the observed frequency of the rules. For a given rule, say $A \rightarrow B$, let $N_{A \rightarrow B}$ be its absolute frequency (count), that is, the number of times at which this rule is observed at least once. In other words, $N_{A \rightarrow B}$ measures the number of transactions in which the rule is satisfied. This does not take into account repeated sequences (occurring more than once), and this may sometimes be a limitation, as in web clickstream analysis. The support for a rule $A \rightarrow B$ is obtained by dividing the number of transactions which satisfy the rule by the total number of transactions, N:

$$\text{support}\{A \rightarrow B\} = \frac{N_{A \rightarrow B}}{N}.$$

The support of a rule is a relative frequency that indicates the proportion of transactions in which the rule is observed. When a large sample is considered, the support approximates the rule's probability of occurrence:

$$\text{support}\{A \rightarrow B\} = \text{Prob}(A \rightarrow B) = \text{Prob } (A \text{ and } B \text{ occur}).$$

The support is quite a useful measure of a rule's interestingness; it is typically employed to filter out rules that are less frequent. The confidence of the rule $A \rightarrow B$ is obtained by dividing the number of transactions which satisfy the rule by the number of transactions which contain the body of the rule, A:

$$\text{confidence}\{A \rightarrow B\} = \frac{N_{A \rightarrow B}}{N_A} = \frac{N_{A \rightarrow B}/N}{N_A/N} = \frac{\text{support}\{A \rightarrow B\}}{\text{support}\{A\}}.$$

The confidence expresses a relative frequency (a probability in the limit) that indicates the proportion of times that, if a rule contains the body A, it will also contain the head B. In other words, it is the frequency (or probability) of occurrence of B, conditionally on A being true. Confidence is the most frequently used interestingness measure of an association rule; it aims to measure the strength of the relationship between two items. For instance, in market basket analysis, the higher the confidence of the association rule $A \rightarrow B$, the greater the probability that if a customer buys products in A, he will also buy product B. In web clickstream analysis, the higher the confidence of the sequence rule $A \rightarrow B$, the greater the probability that if a visitor looks at page A, she will also look at page B.

The language of conditional frequencies and conditional probabilities can be employed to give a normalised strength of the relationship between items A and B. One common measure is the lift; this takes the confidence of a rule and relates it to the support for the rule's head:

$$\text{lift }\{A \rightarrow B\} = \frac{\text{confidence }\{A \rightarrow B\}}{\text{support }\{B\}} = \frac{\text{support}\{A \rightarrow B\}}{\text{support}\{A\} \, \text{support}\{B\}}.$$

Notice how the lift is a ratio between the relative frequency (probability) of both items occurring together, and the relative frequency (probability) of the same event but assuming the two items are independent. Therefore a lift value greater than 1 indicates a positive association, and a value less than 1 a negative association.

These three interestingness measures can be used to search for association rule models in the data. This amounts to finding a set of rules that are statistically interesting. As the number of possible rules is very large, we need some strategies for model selection. One strategy, the forward approach, starts from the simplest rules and proceeds by adding items. This is the approach employed in the well-known Apriori algorithm (Agrawal *et al.*, 1996). From a given set of items, the algorithm starts by selecting a subset for which the support passes a preset threshold t; the other items are discarded. A higher threshold would reduce the complexity of the final solution, as fewer items will be considered.

Next all pairs of items that have passed the previous selection are joined to produce item sets with two items. Item sets are discarded if their support is below the threshold t. The discarded item sets are stored as candidate association rules of order 2; the item selected in this step is the head of the rule. The procedure is repeated. At the mth step, item sets of size m are formed by taking all item sets of size $m - 1$ that have survived and joining them with all those items that have passed the first step. The item sets that do not pass the threshold are discarded and stored to form an association rule of order m; the last item joined is the head and all the previous items are the body. The procedure continues until no rule passes the threshold. The higher the number of variables, with respect to the number of observations, and the higher the threshold value, the quicker the algorithm will terminate.

Notice that the algorithm incorporates a principle of nesting: if a rule of order 2 is discarded, all rules that contain it as antecedent will also be discarded. A disadvantage of the algorithm is that rules with very high confidence or lift, but low support, will not be discovered. Also the algorithm can find rules with high support, high confidence and lift close to 1 (indicating that the two item sets are approximately independent) and flag them as interesting. As the strength of an association is not measured by the support, but by the confidence (or the lift), the Apriori algorithm outputs only those rules that pass a fixed confidence threshold.

An alternative way to generate association rules is by using tree models. This can be seen as an instance of backward search, and is somewhat analogous to pruning in tree models. Indeed, a tree model can be seen as a supervised generator of item sets, each corresponding to the path from the root node to a leaf. In other words, there are as many rules as the tree has leaves. As a tree gives a partition of the observations into exclusive groups, the support and confidence of each decision tree rule can be easily calculated by going through the nodes of the tree. However, the association rules that can be produced by a tree are constructed globally and may be too few and too long. To achieve a larger set of rules, fitted locally, each tree model can be pruned using support and confidence thresholds. The advantage of using a tree representation to construct rules is that pruning is efficient because of their global modelling nature. Furthermore, they can easily deal with all kinds of variables.

The interestingness measures we have used to find rules can also be used to assess the final model (i.e. the list of rules we obtain) by combining the scores of the individual rules. Alternatively, we can use the measures of association introduced in Section 3.4 for analysing interrelationships between qualitative variables. An important difference is that whereas the measures of association refer to all possible pairs of values of the binary variables, association rules consider only the pair (1,1). For instance, as in Section 3.4, the Pearson statistic X^2 is a very general measure of association. It can be used to give an interestingness measure as well:

$$X^2\{A \rightarrow B\} = \frac{(\text{support}\{A \rightarrow B\} - \text{support}\{A\}\,\text{support}\{B\})^2}{\text{support}\{A\}\,\text{support}\{B\}}.$$

This measure of interestingness can be extended to a large number of rules and can be used to assess the departure from independence by appealing to an inferential threshold (based on the chi-square distribution; see Section 4.9). Inferential thresholds can also be derived for association rules. For instance, a large-sample confidence interval for the logarithm of the lift is given by

$$\log(\text{lift}) \pm z_{1-\alpha/2}\sqrt{\frac{1}{\text{support}\{A \rightarrow B\}} - \frac{1}{N} + \frac{1}{\text{support}\{A\}} + \frac{1}{\text{support}\{B\}}},$$

where log(lift) is the observed lift and $z_{1-\alpha/2}$ is the $1 - \alpha/2$ quantile of the standard normal distribution. Exponentiating this expression leads to a confidence interval for the lift. Not only does the width of the interval depend on the confidence level α, it is also directly proportional to the information content of the rule (support$\{A \rightarrow B\}$, support$\{A\}$ and support$\{B\}$) and inversely proportional to the number of transactions N. In other words, the length of the interval, hence the uncertainty on the interestingness of the relationship, decreases as the frequency of the rule increases and in a balanced way (i.e. both the frequency of A and the frequency of B increase).

A confidence interval permits us to decide on the statistical significance of an association rule: if a value of 1 for the lift is within the confidence interval, the rule is not significant. Note that when more than one rule is tested in this way, the conclusions may be overly restrictive, as the tests are not truly independent. In this case it may be appropriate to increase the width of the confidence intervals and therefore reject fewer rules. To assess the validity of a set of rules, we can also use rules based on the comparison between complementary rules, such as $A \rightarrow B$ and $A \rightarrow \overline{B}$, where $\overline{B}$ is the complement of B (true when B is false, and vice versa). A simple one is the odds, seen in Section 3.4:

$$\text{odds}\{A \rightarrow B\} = \frac{\text{support}\{A \rightarrow B\}}{\text{support}\{A \rightarrow \overline{B}\}}.$$

The Gini index and the entropy index can also be applied in this context as measures of heterogeneity for binary variables.

We now consider a specific type of association rule, particularly relevant for some applications. So far we have said that an association rule is simply a rule of joint occurrence between two item sets, A and B. It is possible to attach to this joint occurrence a meaning of logical precedence, so that the body of the rule logically precedes the head of the rule. The resulting rule is called a sequence. Association rules can be specifically calculated for sequences, by linking the transaction data set to an ordering variable. A typical way of introducing a logical precedence is through time. Sequences are not needed in market basket analysis; although products are taken off the shelf in a temporal order, this order is lost when the products are presented at the counter. On the other hand, web clickstream data typically come in as a log file, which preserves the temporal order in which the pages were visited. Therefore it is important to take account of the order in which the pages were visited. When sequences are considered, the meaning of support and confidence changes: support can be interpreted as the number of times that A precedes B; confidence can be interpreted as the conditional probability of B, conditional on A having already occurred.

A further distinction is between direct and indirect sequence rules. A sequence rule is usually indirect, in the sense there may be other elements that logically sit between the body and head of the rule, but they are not considered. For example, if A and B are two web pages, the sequence rule $A \rightarrow B$ searches for all occurrences in which A precedes B, even if other web pages were viewed in between. To allow comparison with the results of global models, it may be interesting to consider direct sequences. A direct sequence searches only for the occurrences in which A exactly precedes B. Note the difference between association and sequence rules. Association rules produce a symmetric relationship, hence the confidence is a measure of association between the binary variables in the two item sets. Sequence rules produce an asymmetric relationship, hence the confidence is a measure of how the variable in the head depends on the variables in the body.

Association rules are probably the best-known local method for detecting relationships between variables. They can be used to mine very large data sets, for which a global analysis may be too complex and unstable. Section 4.14 explains two related types of global model that can provide a very helpful visual representation of the association structures. These models are known as undirected graphical models (for association modelling) and probabilistic expert systems (for dependency modelling). Chapters 7 shows how such global models compare with the local models presented here. Association rules *per se* cannot be used predictively, as there would be more than one sequence to predict a given head of a rule. Tree models can be used predictively and also provide a set of association rules.

As one chapter is entirely devoted to local association models, there are no practical examples in this section. Algorithmic aspects are discussed in Hand *et al.* (2001), which contains a comprehensive description of how to find interesting rules using the Apriori algorithm. The advantages of association rules are their extreme simplicity and interpretational capacity; their disadvantages are the lengthy computing times and analysis costs but, above all, the need for sensible pruning. Software packages produce huge numbers of rules, and without sensible pruning, it is easy to get lost in the details and lose sight of the problem.

4.8.2   Retrieval by content

Retrieval-by-content models are local methods based on identifying a query object of interest then searching the database for the k objects that are most similar to it. In association rules the local aspect is in selecting the variables; in retrieval by content the local aspect is in selecting the observations. The main problem is in finding valid measures of proximity to identify observations that are 'similar'. Notable examples of retrieval by content are searching for information on the internet using a search engine and, more generally, the analysis of text documents, or text mining. The technique is quite broad and can also be applied to audio and video data. There are similarities with memory-based reasoning models (Section 4.7); the main differences are that retrieval by content is not aimed at predicting target variable values, and it is not based on a global assessment of distance between objects, but on distances from the query object. For more details, see Hand *et al.* (2001).

4.9   Uncertainty measures and inference

So far we have not assumed any probabilistic hypothesis of type on the statistical variables of interest. However, the observations considered are generally only a subset from a target population of interest, a sample. Furthermore, the very large size of the data often forces the analyst to consider only a sample of the available data matrix, either for computational reasons (storage and/or processing memory) or for interpretational reasons. Sampling theory gives a statistical explanation of how to sample from a population in order to extract the desired information efficiently; there is not space to cover it here, but Barnett (1974) is a good reference. We shall assume that a sample has been drawn in a random way and is ready for analysis. When dealing with a sample, rather than with the whole population, it is necessary to introduce a probability model that can adequately describe the sampling variability. More generally, a probability model is a useful tool that is often used to model the informational uncertainty that affects most of our everyday decisions.

The introduction of a probability model will lead us to take the estimated statistical summaries and attach measures of variability that describe the degree of uncertainty in the estimate due to sample variability. This will eventually lead us to substitute parametric point estimates with so-called interval estimates; we replace a number with an interval of numbers that contains the parameter of interest in most cases. We can improve the diagnostic ability of a model by using statistical hypothesis testing; for example, we can introduce a critical threshold above which we retain a certain regression plane as a valid description of the relationship between the variables or we treat a certain clustering of the data as a valid partitioning of the observations. For descriptions of the various probability models, see Mood *et al.* (1991) or Bickel and Doksum (1977).

4.9.1 Probability

An event is any proposition that can be either true or false and is formally a subset of the space Ω, which is called the space of all elementary events. Elementary events are events that cannot be further decomposed, and cover all possible occurrences. Let $\mathfrak{F}$ be a class of subsets of Ω, called the event space. A probability function P is a function defined on $\mathfrak{F}$ that satisfies the following axioms:

- $P(A) \geq 0, \ \forall A \in \mathfrak{F}$.
- $P(\Omega) = 1$.
- If $A_1, A_2, \ldots$ is a sequence of events of $\mathfrak{F}$ pairwise mutually exclusive (i.e. $A_i \cap A_j = \emptyset$ for $i \neq j$, $i, j = 1, 2, \ldots$) and if $A_1 \cup A_2 \cup \ldots = \bigcup_{i=1}^{\infty} A_i \in \mathfrak{F}$, then $P\left(\bigcup_{i=1}^{\infty} A_i\right) = \sum_{i=1}^{\infty} P(A_i)$.

A probability function will also be referred to as a probability measure or simply probability. The three axioms can be interpreted in the following way. The first says that the probability is a non-negative function. The second says that the probability of the event Ω is 1; Ω is an event that will always be true as it coincides with all possible occurrences. Since any event is a subset of Ω, it follows that the probability of any event is a real number in [0,1]. The third axiom says that the probability of occurrence of any one of a collection of events (possibly infinite, and mutually exclusive) is the sum of the probabilities of occurrence of each of them. This is the formal, axiomatic definition of probability due to Kolmogorov (1933). There are several interpretations of this probability. These interpretations will help us from an operational viewpoint when we come to construct a probability measure. In the classical interpretation, if an experiment gives rise to a finite number n of possible results, then $P(A) = n_A/n$, where n_A denotes the number of results in A (favourable results). In the more general frequentist interpretation, the probability of an event coincides with the relative frequency of the same event in a large (possibly infinite) sequence of repeated trials in the same experimental conditions. The frequentist interpretation allows us to take most of the concepts developed for frequencies (such as those in Chapter 3) and extend them to the realm of probabilities. In the even more general (although somewhat controversial) subjective interpretation, probability is a degree of belief that an individual attaches to the occurrence of an event. This degree of belief is totally subjective but not arbitrary, since probabilities must obey coherency rules, corresponding to the above axioms and all the rules derivable from them. The advantage of the subjective approach is that it is always applicable, especially when an event cannot be repeated (a typical situation for observational data and data mining, and unlike experimental data).

We can use the three axioms to deduce the basic rules of probability can be deduced. Here are the complement rule and the union rule.

- **Complement rule.** If A is any event in $\mathfrak{F}$, and $\overline{A}$ its complement (negation), then $P(\overline{A}) = 1 - P(A)$.
- **Union rule.** For any pair of events A, $B \in \mathfrak{F}$, $P(A \cup B) = P(A) + P(B) - P(A \cap B)$, where the union event, $A \cup B$, is true when either A or B is true; while the intersection event $A \cap B$ is true when both A and B are true.

Probability has so far been defined in the absence of information. Similar to the concept of relative frequency, we can define the probability of an event A occurring, conditional on the information that the event B is true. Let A and B be two events in $\mathfrak{F}$. The conditional probability of the event A, given that B is true, is

$$P(A|B) = \frac{P(A \cap B)}{P(B)}, \text{ with } P(B) > 0.$$

This definition extends to any conditioning sets of events. Conditional probabilities allow us to introduce further important rules.

- **Intersection rule.** Let A and B two events in $\mathfrak{F}$. Then $P(A \cap B) = P(A|B)P(B) = P(B|A)P(A)$.
- **Independence of events.** If A is independent of B, then

$$P(A \cap B) = P(A)P(B),$$

$$P(A|B) = P(A),$$

$$P(B|A) = P(B).$$

In other words, if two events are independent, knowing that one of them occurs does not alter the probability that the other one occurs.

- **Total probability rule.** Consider n events H_i, $i = 1, \ldots, n$, pairwise mutually exclusive and exhaustive, in Ω (equivalently, they form a partition of Ω), with $P(H_i) > 0$. Then the probability of an event B in $\mathfrak{F}$ is given by

$$P(B) = \sum_{i=1}^{n} P(B|H_i)P(H_i).$$

- **Bayes' rule.** Consider n events H_i, $i = 1, \ldots, n$, pairwise mutually exclusive and exhaustive, in Ω (equivalently, they form a partition of Ω), with $P(H_i) > 0$. Then the probability of an event B in $\mathfrak{F}$ such that $P(B) > 0$ is given by

$$P(H_i|B) = \frac{P(B|H_i)P(H_i)}{\sum_j P(B|H_j)P(H_j)}.$$

The total probability rule plays a very important role in the combination of different probability statements; we will see an important application later. Bayes' rule is also very important; also known as the 'inversion rule', it calculates the conditional probability of an event by using the reversed conditional probabilities. Note also that the denominator of Bayes' rule is the result of the total probability rule; it acts as a normalising constant of the probabilities in the numerator. This theorem lies at the heart of the inferential methodology known as Bayesian statistics.

4.9.2 Statistical models

Suppose that, for the problem at hand, we have defined all the possible elementary events Ω, as well as the event space $\mathfrak{F}$. Suppose also that, on the basis of one of the operational notions of probability, we have constructed a probability measure, P. The triplet $(\Omega, \mathfrak{F}, P)$ defines a probability space; is is the basic ingredient for defining a random variable, hence for constructing a statistical model.

Given a probability space $(\Omega, \mathfrak{F}, P)$, a random variable is any function $X(\omega)$, $\omega \in \Omega$, with values on the real line. The cumulative distribution of a random variable X, denoted by F, is a function defined on the real line, with values on [0,1], that satisfies $F(x) = P(X \le x)$ for any real number x. The cumulative distribution function (often called the distribution function for short) characterises the probability distribution for X. It is the main tool for defining a statistical model of the uncertainty concerning a variable X.

We now examine two special important cases of random variables, and look at their distribution functions. A random variable is said to be discrete if it can take only a finite, or countable, set of values. In this case

$$F(x) = \sum_{X \le x} p(x), \text{ with } p(x) = P(X = x).$$

Therefore in this case $p(x)$, called the discrete probability function, also characterises the distribution. Note that both quantitative discrete and qualitative variables can be modelled with a discrete random variable, provided that numerical codes are assigned to qualitative variables. They are collectively known as categorical random variables.

A random variable is said to be continuous if there exists a function f, called the density function, such that the distribution function can be obtained from it:

$$F(x) = \int_{-\infty}^{x} f(u)du, \text{ for any real number } x.$$

Furthermore, the density function has the following two properties:

$$f(x) \ge 0, \qquad \forall x,$$

$$\int_{-\infty}^{\infty} f(x)dx = 1.$$

In view of its definition, the density function characterises a statistical model for continuous random variables.

By replacing relative frequencies with probabilities, we can treat random variables like the statistical variables seen in Chapter 3. For instance, the discrete probability function can be taken as the limiting relative frequency of a discrete random variable. On the other hand, the density function corresponds to the height of the histogram of a continuous variable. Consequently, the concepts in Chapter 3 – mean, variance, correlation, association, etc. – carry over to random variables. For instance, the mean of a random variable, usually called the expected value, is defined by

$$\mu = \begin{cases} \sum x_i p_i & \text{if } X \text{ is categorical,} \\ \int x f(x) dx & \text{if } X \text{ is continuous.} \end{cases}$$

The concept of a random variable can be extended to cover random vectors or other random elements, thereby defining a more complex statistical model. Henceforth, we use notation for random variables, without loss of generality.

In general, a statistical model of uncertainty can be defined by the pair $(X, F(x))$, where X is a random variable and $F(x)$ is the cumulative distribution attached to it. It is often convenient to specify I directly, choosing it from a catalogue of models available in the statistical literature, models which have been constructed specifically for certain problems. These models can be divided into three main classes: parametric models, for which the cumulative distribution is completely specified by a finite set of parameters, denoted by θ; non-parametric models, which require the whole specification of F; and semiparametric models, where the specification of F is eased by having some parameters but these parameters do not fully specify the model.

We now examine the most frequently used parametric model, the Gaussian distribution; Section 4.10 looks at non-parametric and semiparametric models. Let Z be a continuous variable, with real values. Z is distributed according to a standardised Gaussian (or normal) distribution if the density function is

$$f(z) = \frac{1}{\sqrt{2\pi}} e^{-z^2/2}.$$

This is a bell-shaped distribution (Section 3.1), with most of the probability around its centre, which coincides with the mean, the mode and the median of the distribution (equal to 0 for the standardised Gaussian distribution). Since the distribution is symmetric, the probability of having a value greater than a certain positive quantity is equal to the probability of having a value lower than the negative of the same quantity – for example, $P(Z > 2) = P(Z < -2)$. Having defined the Gaussian as our reference model, we can use it to calculate some probabilities of interest; these probabilities are areas under the density function. We cannot calculate them in closed form, so we must use numerical approximation. In the past this involved the use of statistical tables but now it can be done with all the main data analysis packages. Here is a financial example.

Consider the valuation of the return on a certain financial activity. Suppose, as is often done in practice, that the future distribution of this return, Z, expressed in euros, follows the standardised Gaussian distribution. What is the probability of observing a return greater than €1? To solve this problem it is sufficient to calculate the probability $P(Z\rangle 1)$. The solution is not expressible in closed form, but using statistical software we find that the probability is equal to about 0.159.

Now suppose that a financial institution has to allocate an amount of capital to be protected against the risk of facing a loss on a certain portfolio. This problem is a simplified version of a problem faced by credit operators every day – calculating value at risk (VaR). VaR is a statistical index that measures the maximum loss to which a portfolio is exposed in a holding period Δt and with a fixed level α of desired risk. Let Z be the change in value of the portfolio during the period considered, expressed in standardised terms. The VaR of the portfolio is then the loss (corresponding to a negative return), implicitly defined by

$$P(Z \leq -\text{VaR}) = 1 - \alpha.$$

Suppose that the level of desired risk is 5%. This corresponds to fixing the right-hand side at 0.95; the value of the area under the standardised density curve to the right of the value VaR (i.e. to the right of the value – VaR) is then equal to 0.05. Therefore, the VaR is given by the point on the x-axis of the graph that corresponds to this area. The equation has no closed-form solution. But statistical software easily computes compute that VaR = 1.64. Figure 4.9 illustrates the calculation of the VaR. The histogram shows the observed returns and the continuous line is the standard Gaussian distribution, used to calculate the VaR. In quantitative risk management this approach is known as the analytic approach or the delta normal approach, in contrast to simulation-based methods.

So far we have considered the standardised Gaussian distribution, with mean 0 and variance 1. It is possible to obtain a family of Gaussian distributions that differ only in their mean and variance. In other words, the Gaussian distribution is a parametric statistical model, parameterised by two parameters. Formally, if Z is a standard Gaussian random variable and $X = \sigma Z + \mu$ then X is distributed according to a Gaussian distribution with mean μ and variance $\sigma 2$. The family of Gaussian distributions is closed with respect to linear transformations; that is, any linear transformation of a Gaussian variable is also Gaussian. As a result, the Gaussian distribution is well suited to situations in which we hypothesise linear relationships among variables.

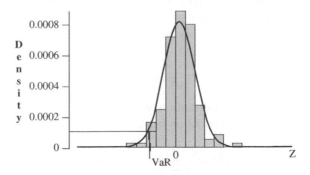

Figure 4.9   Calculation of VaR.

Our definition of the Gaussian distribution can be extended to the multivariate case. The resulting distribution is the main statistical model for the inferential analysis of continuous random vectors. For simplicity, here is the bivariate case. A two-dimensional random vector (X_1, X_2) has a bivariate Gaussian distribution if there exist six real constants,

$$a_{ij}, \quad 1 \le i, j \le 2,$$
$$\mu_i, \quad i = 1, 2,$$

and two independent standardised Gaussian random variables, Z_1 and Z_2, such that

$$X_1 = \mu_1 + a_{11}Z_1 + a_{12}Z_2,$$
$$X_2 = \mu_2 + a_{21}Z_1 + a_{22}Z_2.$$

In matrix terms, the previous equation can be written as $\mathbf{X} = \boldsymbol{\mu} + \mathbf{AZ}$, which easily extends to the multivariate case. In general, therefore, a multivariate normal distribution is completely specified by two parameters, the mean vector $\boldsymbol{\mu}$ and the variance–covariance matrix $\boldsymbol{\Sigma} = \mathbf{AA}'$.

Using the Gaussian distribution, we can derive three distributions of special importance in inferential analysis: the chi-squared distribution, the Student's t distribution and the F distribution.

The chi-squared distribution is obtained from a standardised Gaussian distribution. If Z is a standardised Gaussian distribution, the random variable defined by Z^2 is said to follow a chi-squared distribution, with 1 degree of freedom, denoted by $\chi^2(1)$. More generally, a parametric family of chi-squared distributions, indexed by one parameter, is obtained from the fact that the sum of n independent chi-squared distributions with 1 degree of freedom is a chi-squared with n degrees of freedom, $\chi^2(n)$. The chi-squared distribution has positive density only for positive real values. Probabilities from it have to be calculated numerically, as for the Gaussian distribution. Finally, the chi-squared value has an expected value equal to n and a variance equal to $2n$.

The Student's t distribution is characterised by a density symmetric around zero, like the Gaussian but more peaked (i.e. with a higher kurtosis). It is described by one parameter, the degrees of freedom, n. As n increases, the Student's t distribution approaches the Gaussian distribution. Formally, let Z be a standardised Gaussian distributions, in symbols $Z \sim N(0,1)$, and let U be a chi-squared distribution, with n degrees of freedom, $U \sim \chi_n^2$. If Z and U are independent, then

$$T = \frac{Z}{\sqrt{U/n}} \sim t(n),$$

that is, T is a Student's t distribution with n degrees of freedom. It can be shown that the Student's t distribution has an expected value of 0 and a variance given by

$$\mathrm{Var}(T) = \frac{n}{n-2}, \qquad \text{for } n > 2.$$

Finally, the F distribution is also asymmetric and defined only for positive values, like the chi-squared distribution. It is obtained as the ratio between two independent chi-squared distributions, U and V, with degrees of freedom m and n, respectively:

$$F = \frac{U/m}{V/n}.$$

The F distribution is therefore described by two parameters, m and n; it has an expected value equal to $n/(n-2)$ and a variance which is a function of both m and n. An F distribution with $m = 1$ is equal to the square of a Student's t with n degrees of freedom.

4.9.3 Statistical inference

Statistical inference is mainly concerned with the induction of general statements on a population of interest, on the basis of the observed sample. First we need to first derive the expression for the distribution function for a sample of observations from a statistical model. A sample of n observations on a random variable X is a sequence of random variables $X_1, X_2, \ldots, X_n$ that are distributed identically to X. In most cases it is convenient to assume that the sample is a simple random sample, with the observations drawn with replacement from the population modelled by X. Then it follows that the random variables $X_1, X_2, \ldots, X_n$ are independent and therefore constitute a sequence of independent and identically distributed (i.i.d.) random variables. Let $\mathbf{X}$ denote the random vector formed by such a sequence of random variables, $\mathbf{X} = (X_1, X_2, \ldots, X_n)$, and $\mathbf{x} = (x_1, x_2, \ldots, x_n)$ denote the sample value actually observed. It can be shown that, if the observations are i.i.d., the cumulative distribution of $\mathbf{X}$ simplifies to

$$F(\mathbf{x}) = \prod_{i=1}^{n} F(x_i),$$

with $F(x_i)$ the cumulative distribution of X, evaluated for each of the sample values $(x_1, x_2, \ldots, x_n)$. If $\mathbf{x} = (x_1, x_2, \ldots, x_n)$ are the observed sample values, this expression gives a probability, according to the assumed statistical model, of observing sample values less than or equal to the observed vales. Furthermore, when X is a continuous random variable,

$$f(\mathbf{x}) = \prod_{i=1}^{n} f(x_i),$$

where f is the density function of X. And when X is a discrete random variable,

$$p(\mathbf{x}) = \prod_{i=1}^{n} p(x_i),$$

where p is the discrete probability function of X. If $\mathbf{x} = (x_1, x_2, \ldots, x_n)$ are the observed sample values, this expression gives the probability, according to the assumed statistical model, of observing sample values exactly equal to the observed vales. In other words, it measures how good the assumed model is for the given data. A high value of $p(\mathbf{x})$, possibly close to 1, implies that the data are well described by the statistical model; a low value of $p(\mathbf{x})$ implies that the data are poorly described. Similar conclusions can be drawn for $f(\mathbf{x})$ in the continuous case. The difference is that the sample density $f(\mathbf{x})$ is not constrained to be in [0,1], unlike the sample probability $p(\mathbf{x})$. Nevertheless, higher values of $f(\mathbf{x})$ also indicate that the data are well described by the model, and low values indicate that the data are poorly described. In both cases we can say that $p(\mathbf{x})$ or $f(\mathbf{x})$ express the *likelihood* of the model for the given data.

These are fundamentals ideas when considering inference. A statistical model is a rather general model, in the sense that once a model is assumed to hold, it remains to specify precisely the distribution function or, if the model is parametric, the unknown parameters. In general, there remain unknown quantities to be specified. This can seldom be done theoretically, without reference to the observed data. As the observed data are typically observed on a sample, the main purpose of statistical inference is to 'extend' the validity of the calculations obtained on the sample to the whole population. In this respect, when statistical summaries are calculated on a sample rather than a whole population, it is more correct to use the term 'estimated' rather than 'calculated', to reflect the fact that the values obtained depend on the sample chosen and may therefore be different if a different sample is considered. The summary functions that produce the estimates, when applied to the data, are called statistics. The simplest examples of statistics are the sample mean and the sample variance; other examples are the statistical indexes in Chapter 3, when calculated on a sample.

The methods of statistical inference can be divided into estimation methods and hypothesis testing procedures. Estimation methods derive statistics, called estimators, of the unknown model quantities that, when applied to the sample data, can produce reliable estimates of them. Estimation methods can be divided into point estimate methods, where the quantity is estimated with a precise value, and confidence interval methods, where the quantity is estimated to have a high frequency of lying within a region, usually an interval of the real line. To provide a confidence interval, estimators are usually supplemented by measures of their sampling variability. Hypothesis testing procedures look at the use of the statistics to take decisions and actions. More precisely, the chosen statistics are used to accept or reject a hypothesis about the unknown quantities by constructing useful rejection regions that impose thresholds on the values of the statistics.

We briefly present the most important inferential methods. For simplicity, we refer to a parametric model. Starting with estimation methods, consider some desirable properties for an estimator. An estimator T is said to be unbiased, for a parameter θ, if $E(T) = \theta$. The difference $E(T) - \theta$ is called bias of an estimator and is null if the estimator is unbiased. For example, the sample mean

$$\overline{X} = \frac{1}{n} \sum_{i=1}^{n} X_i$$

is always an unbiased estimator of the unknown population mean μ, as it can be shown that $E(\overline{X}) = \mu$. On the other hand, the sample variance

$$S^2 = \frac{1}{n} \sum_{i=1}^{n} (X_i - \overline{X})^2$$

is a biased estimator of the sample variance σ^2, as

$$E(S^2) = \frac{n-1}{n} \sigma^2.$$

Its bias is therefore

$$\text{bias}(S^2) = -\frac{1}{n} \sigma^2.$$

This explains why an often used estimator of the population variance is the unbiased sample variance

$$S^2 = \frac{1}{n-1} \sum_{i=1}^{n} (X_i - \overline{X})^2.$$

A related concept is the efficiency of an estimator, which is a relative concept. Among a class of estimators, the most efficient estimator is usually the one with the lowest mean squared error (MSE), which is defined on the basis of the Euclidean distance by

$$\text{MSE}(T) = E[(T - \theta)^2].$$

It can be shown that

$$\text{MSE}(T) = [\text{bias}(T)]^2 + \text{Var}(T).$$

MSE thus has two components: the bias and the variance. As we shall see in Chapter 5, there is usually a trade-off between these quantities: if one increases, the other decreases. The sample mean can be shown to be the most efficient estimator of the population mean. For large samples, this can be easily seen by applying the definition.

Finally, an estimator is said to be consistent (in quadratic mean) if, for $n \to \infty$, $\lim(\text{MSE}(T)) = 0$. This implies that, for $n \to \infty$, $P(\lim |T - \theta|) = 1$; that is, for a large sample, the probability that the estimator lies in an arbitrarily small neighbourhood of θ approximates to 1. Notice that both the sample mean and the sample variances introduced above are consistent.

In practice, the two most important estimation methods are the maximum likelihood and Bayesian methods.

Maximum likelihood methods

Maximum likelihood methods start by considering the likelihood of a model which, in the parametric case, is the joint density of $\mathbf{X}$, expressed as a function of the unknown parameters θ:

$$p(\mathbf{x}; \theta) = \prod_{i=1}^{n} p(x_i; \theta),$$

where θ are the unknown parameters and $\mathbf{X}$ is assumed to be discrete.

The same expression holds for the continuous case, with p replaced by f. In the rest of the text we will therefore use the discrete notation, without loss of generality, case.

If a parametric model is chosen, the model is assumed to have a precise form, and the only unknown quantities left are the parameters. Therefore the likelihood is in fact a function of the parameters θ. To stress this fact, the previous expression can be also denoted by $L(\theta; \mathbf{x})$. Maximum likelihood methods suggest that, as estimators of the unknown parameter θ, we take the statistics that maximise $L(\theta; \mathbf{x})$ with respect to θ. The heuristic motivation for maximum likelihood is that it selects the parameter value that makes the observed data most likely under the assumed statistical model. The statistics generated using maximum likelihood are known as maximum likelihood estimators (MLEs) and have many desirable properties. In particular, they can be used to derive confidence intervals. The typical procedure is to assume that a large sample is available (this is often the case in data mining), in which case the MLE is approximately distributed as a Gaussian (normal) distribution. The estimator can thus be used in a simple way to derive an asymptotic (valid for large samples) confidence interval. For example, let T be an MLE and let $\mathrm{Var}(T)$ be its asymptotic variance. Then a $100\,(1 - \theta)\%$ confidence interval is given by

$$\left(T - z_{1-\alpha/2}\sqrt{\mathrm{Var}(T)}, \; T + z_{1-\alpha/2}\sqrt{\mathrm{Var}(T)} \right),$$

where $z_{1-\alpha/2}$ is the $100(1 - \alpha/2)$ percentile of the standardised normal distribution, such that the probability of obtaining a value less than $z_{1-\alpha/2}$ is equal to $1 - \alpha/2$. The quantity $1 - \alpha$ is also known as the confidence level of the interval, as it gives the confidence that the procedure is correct: in $100(1 - \alpha)\%$ of cases the unknown quantity will fall within the chosen interval. It has to be specified before the analysis. For the normal distribution, the estimator of μ is the sample mean, $\overline{X} = n^{-1} \sum X_i$. So a confidence interval for the mean, when the variance σ^2 is assumed to be known, is given by

$$\left(\overline{X} - z_{1-\alpha/2}\sqrt{\mathrm{Var}(\overline{X})}, \; \overline{X} + z_{1-\alpha/2}\sqrt{\mathrm{Var}(\overline{X})} \right), \qquad \text{where } \mathrm{Var}(\overline{X}) = \frac{\sigma^2}{n}.$$

When the distribution is normal from the start, as in this case, the expression for the confidence interval holds for any sample size. A common procedure in confidence intervals is to assume a confidence level of 95%; in a normal distribution this leads to $z_{1-\alpha/2} = 1.96$.

Bayesian methods

Bayesian methods use Bayes' rule, which provides a powerful framework for combining sample information with (prior) expert opinion to produce an updated (posterior) expert opinion. In Bayesian analysis, a parameter is treated as a random variable whose uncertainty is modelled by a probability distribution. This distribution is the expert's prior distribution $p(\theta)$, stated in the absence of the sampled data. The likelihood is the distribution of the sample, conditional on the values of the random variable θ, $p(\mathbf{x}|\theta)$. Bayes' rule provides an algorithm to update the expert's opinion in light of the data, producing the so-called posterior distribution $p(\theta|\mathbf{x})$:

$$p(\theta|\mathbf{x}) = c^{-1} p(\mathbf{x}|\theta) p(\theta),$$

with $c = p(\mathbf{x})$, a constant that does not depend on the unknown parameter θ. The posterior distribution represents the main tool of Bayesian inference. Once it is obtained, it is easy to obtain any inference of interest. For instance, to obtain a point estimate, we can take a summary of the posterior distribution, such as the mean or the mode. Similarly, confidence intervals can be easily derived by taking any two values of θ such that the probability of θ belonging to the interval described by those two values corresponds to the given confidence level. As θ is a random variable, it is now correct to interpret the confidence level as a probability statement: $1 - \alpha$ is the coverage probability of the interval, namely, the probability that θ assumes values in the interval. The Bayesian approach is thus a coherent and flexible procedure. On the other hand, it has the disadvantage of requiring a more computationally intensive approach, as well as more careful statistical thinking, especially in providing an appropriate prior distribution.

For the normal distribution example, assuming as a prior distribution for θ a constant distribution (expressing a vague state of prior knowledge), the posterior mode is equal to the MLE. Therefore, maximum likelihood estimates can be seen as a special case of Bayesian estimates. More generally, it can be shown that, when a large sample is considered, the Bayesian posterior distribution approaches an asymptotic normal distribution, with the maximum likelihood estimate as expected value. This reinforces the previous conclusion.

An important application of Bayes' rule arises in predictive classification problems. As explained in Section 4.4, the discriminant rule establishes that an observation x is allocated to the class with the highest probability of occurrence, on the basis of the observed data. This can be stated more precisely by appealing to Bayes' rule. Let C_i, for $i = 1, \ldots, k$, denote a partition of mutually exclusive and exhaustive classes. Bayes' discriminant rule allocates each observation x to the class C_i that maximises the posterior probability:

$$p(C_i|x) = c^{-1} p(x|C_i) p(C_i), \qquad \text{where } c = p(x) = \sum_{i=1}^{k} p(C_i) p(x|C_i),$$

and x is the observed sample. Since the denominator does not depend on C_i, it is sufficient to maximise the numerator. If the prior class probabilities are all equal to k^{-1}, maximisation of the posterior probability is equivalent to maximisation of the likelihood $p(x|C_i)$. This is the approach often followed in practice. Another common approach is to estimate the prior class probabilities with the observed relative frequencies in a training sample. In any case, it can be shown that the Bayes discriminant rule is optimal, in the sense that it leads to the least possible misclassification error rate. This error rate is measured as the expected error probability when each observation is classified according to Bayes' rule:

$$p_B = \int \left(1 - \max_i p(C_i|x)\right) p(x) dx,$$

also known as the Bayes error rate. No other discriminant rule can do better than a Bayes classifier; that is, the Bayes error rate is a lower bound on the misclassification rates. Only rules that derive from Bayes' rule are optimal. For instance, the logistic discriminant rule and the linear discriminant rule (Section 4.4) are optimal, whereas the discriminant rules obtained from tree models, multilayer perceptrons and nearest-neighbour models are optimal for a large sample size.

Hypothesis testing

We now briefly consider procedures for hypothesis testing. A statistical hypothesis is an assertion about an unknown population quantity. Hypothesis testing is generally performed in a pairwise way: a null hypothesis H_0 specifies the hypothesis to be verified, and an alternative hypothesis H_1 specifies the hypothesis with which to compare it. A hypothesis testing procedure is usually constructed by finding a rejection (critical) rule such that H_0 is rejected, when an observed sample statistic satisfies that rule, and vice versa. The simplest way to construct a rejection rule is by using confidence intervals. Let the acceptance region of a test be defined as the logical complement of the rejection region. An acceptance region for a (two-sided) hypothesis can be obtained from the two inequalities describing a confidence interval, swapping the parameter with the statistic and setting the parameter value equal to the null hypothesis. The rejection region is finally obtained by inverting the signs of the inequalities. For instance, in our normal distribution example, the hypothesis H_0: $\mu = 0$ will be rejected against the alternative hypothesis H_1: $\mu \neq 0$ when the observed value of $\overline{X}$ is outside the interval

$$\left(0 - z_{1-\alpha/2}\sqrt{\mathrm{Var}(\overline{X})}, 0 + z_{1-\alpha/2}\sqrt{\mathrm{Var}(\overline{X})}\right).$$

The probability α has to be specified a priori and is called the significance level of the procedure. It corresponds to the probability of a type I error, that is, the probability of rejecting the null hypothesis when it is actually true. A common assumption is to take $\alpha = 0.05$, which corresponds to a confidence level of 0.95. The probability is obtained, in this case, by summing two probabilities relative to the random variable $\overline{X}$: the probability that $\overline{X} < 0 - z_{1-\alpha/2}\sqrt{\mathrm{Var}(\overline{X})}$ and the probability that $\overline{X} > 0 + z_{1-\alpha/2}\sqrt{\mathrm{Var}(\overline{X})}$. Notice that the rejection region is derived by setting $\mu = 0$. The significance level is calculated using the same assumption. These are general facts: statistical tests are usually derived under the assumption that the null hypothesis is true. This is expressed by saying that the test holds 'under the null hypothesis'. The limits of the interval are known as critical values. If the alternative hypothesis were one-sided, the rejection region would correspond to only one inequality. For example, if $H_1: \mu > 0$, the rejection region would be defined by the inequality $\overline{X} > 0 + z_{1-\alpha}\sqrt{\mathrm{Var}(\overline{X})}$. The critical value is different because the significance level is now obtained by considering only one probability.

There are other methods for deriving rejection rules. An alternative approach to testing the validity of a certain null hypothesis is by calculating the p-value of the test. The p-value can be described as the probability, calculated under the null hypothesis, of observing a test statistic more extreme than actually observed, assuming the null hypothesis is true, where 'more extreme' means in the direction of the alternative hypothesis. For a two-sided hypothesis, the p-value is usually taken to be twice the one-sided p-value. In our normal distribution example, the test statistic is $\overline{X}$. Let $\bar{x}$ be the observed sample value of $\overline{X}$. The p-value would then be equal to twice the probability that $\overline{X}$ is greater than $\bar{x}$: p-value $= 2P(X > x)$. A small p-value will indicate that $\bar{x}$ is far from the null hypothesis, which is thus rejected; a large p-value will mean that the null hypothesis cannot be rejected. The threshold value is usually the significance level of the test, which is chosen in advance. For instance, if the chosen significance level of the test is $\alpha = 0.05$, a p-value of 0.03 indicates that the null hypothesis can be rejected, whereas a p-value of 0.18 indicates that the null hypothesis cannot be rejected.

4.10 Non-parametric modelling

A parametric model is usually specified by making a hypothesis about the distribution and by assuming this hypothesis is true. But this can often be difficult or uncertain. One possible way to overcome this is to use non-parametric procedures, which eliminate the need to specify the form of the distribution in advance. A non-parametric model only assumes that the observations come from a certain distribution function F, not specified by any parameters. But compared with parametric models, non-parametric models are more difficult to interpret and estimate. Semiparametric models are a compromise between parametric models and non-parametric models.

A non-parametric model can be characterised by the distribution function or by the density function, which need to be fully specified. First consider the estimate of the distribution function. A valid estimator is the empirical distribution function, usually denoted by $S(x)$. Intuitively it is an analogous estimate of the distribution function $F(x)$ of the random variable X. Formally, the empirical distribution function is calculated, at any point x, by taking the proportion of sample observations less or equal to it,

$$S(x) = \frac{1}{n}\#\{x_i \leq x\}.$$

It can be shown that the expected value of $S(x)$ is $F(x)$ and that

$$\text{Var}(S(x)) = \frac{1}{n} F(x)(1 - F(x)).$$

Therefore the empirical distribution function is an unbiased estimator of $F(x)$ and it is consistent as, for $n \to \infty$, $\text{Var}(S(x)) \to 0$, so that $\text{MSE}(S(x)) \to 0$.

The sample distribution function can be used to assess a parametric model's goodness of fit in an exploratory way. To evaluate the goodness of fit of a distribution function, we usually use the Kolmogorov–Smirnov distance that leads to the well-known statistical test of the same name. In this test the null hypothesis refers to a particular distribution which we shall call $F^*(x)$ (this distribution could be Gaussian, for example). Therefore we have

$$H_0 : F(x) = F^*(x),$$

$$H_1 : F(x) \neq F^*(x).$$

To test H_0 against H_1 we consider the available random sample $X_1, \ldots, X_n$. The idea is to compare the observed distribution function, $S(x)$, with the theoretical distribution function F^* calculated with the observed values. The idea of Kolmogorov and Smirnov is simple and clever. Since $S(x)$ estimates $F(x)$ it is logical to hypothesise a 'distance'" between $S(x)$ and $F(x)$. If $S(x)$ and $F(x)$ are close enough (i.e. similar enough) the null hypothesis can be accepted, otherwise it is rejected. But what kind of test statistics can we use to measure the discrepancy between $S(x)$ and $F(x)$? One of the easiest measurements is the supremum of the vertical distance between the two functions. This is the statistic suggested by Kolmogorov:

$$T_1 = \sup_{-\infty < x < +\infty} \left| S(x) - F^*(x) \right|.$$

It relies on the usage of the uniform distance, explained in Section 5.1. For 'high' values of T_1, the null hypothesis is rejected, while for 'low' values it is accepted. The logic of the T_1 statistic is obvious but the calculation of the probability distribution is more complicated. Nevertheless, we can demonstrate that, under the null hypothesis, the probability distribution of the statistical test based on T_1 does not depend on the functional form of $F^*(x)$. This distribution is tabulated and included in the main statistical packages. It is therefore possible to determine critical values for T_1 and obtain a rejection region of the null hypotheses. Alternatively, it is possible to obtain p-values for the test. The Kolmogorov–Smirnov test is important in exploratory analysis. For example, when the QQ plot (Section 3.1) does not give any obvious indications that a certain empirical distribution is normal or not, we can check whether the distance of the normal distribution function from the empirical distribution function is large enough to be rejected. Figure 4.10 illustrates how the Kolmogorov–Smirnov statistic works.

The simplest type of density estimator is the histogram. A histogram assigns a constant density to each interval class. This density is easily calculated by taking the relative frequency of observations in the class and dividing it by the class width. For continuous densities, the histogram can be interpolated by joining

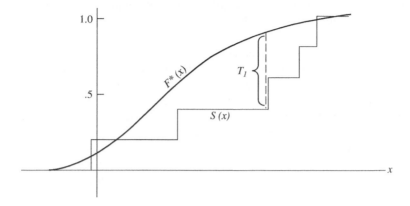

Figure 4.10   The Kolmogorov–Smirnov statistic.

all midpoints of the top segment of each bar. However, histograms can depend heavily on the choice of the classes, as well as on the sample, especially when considering a small sample. Kernel estimators represent a more refined class of density estimators. They represent a descriptive model, that however works locally, strongly analogous to nearest-neighbour models (Section 4.7). Consider a continuous random variable X, with observed values $x_1, \ldots, x_n$, and a kernel density function K with a bandwidth h. The estimated density function at any point x is

$$\hat{f}(x) = \frac{1}{n} \sum_{i=1}^{n} K\left(\frac{x - x_i}{h}\right).$$

In practice the kernel function is usually chosen as a unimodal function, with a mode at zero. A common choice is to take a normal distribution for the random variable $x - x_i$ with zero mean and variance corresponding to h^2, the square of the bandwidth of the distribution. The quality of a kernel estimate then depends on a good choice of the variance parameter h. The choice of h reflects the trade-off between parsimony and goodness of fit that we have already encountered: a low value of h is such that the estimated density values are fitted very locally, possibly on the basis of a single data point; a high value leads to a global estimate, smoothing the data too much. It is quite difficult to establish what a good value of h should be. One possibility is to use computationally intensive methods, such as cross-validation techniques. The training sample is used to fit the density, and the validation sample to calculate the likelihood of the estimated density. A value of h can then be chosen that leads to a high likelihood.

Estimating high-dimensional density functions is more difficult but kernel methods can still be applied. Replacing the univariate normal kernel with a multivariate normal kernel yields a viable multivariate density estimator. Another approach is to assume that the joint density is the product of univariate kernels. However, the problem is that, as the number of variables increases, observations tend to be farther away and there is little data for the bandwidths. This parallels what happens with nearest-neighbour models. Indeed, both are memory-based and the main difference is in their goals; kernel models are descriptive and nearest-neighbour models are predictive.

Kernel methods can be seen as a useful model for summarising a low-dimensional data set in a non-parametric way. This can be a helpful step towards the construction of a parametric model, for instance.

The most important semiparametric models are mixture models. These models are suited to situations where the data set can be clustered into groups of obser-vations, each with a different parametric form. The model is semiparametric because the number of groups, hence the number of distributions to consider, is unknown. The general form of a finite mixture distribution for a random variable X is

$$f(x) = \sum_{i=1}^{g} w_i f_i(x; \theta_i),$$

where w_i is the probability that an observation is distributed as the ith popula-tion, with density f_i and parameter vector θ_i. Usually the density functions are all the same (often normal) and this simplifies the analysis. We can apply a sim-ilar techniques to a random vector X. The model can be used for (model-based) probabilistic cluster analysis. Its advantage is conducting cluster analysis in a coherent probability framework, allowing us to draw conclusions based on infer-ential results rather than on heuristics. Its disadvantage is that the procedure is structurally complex and possibly time-consuming. The model can choose the number of components (clusters) and estimate the parameters of each population as well as the weight probabilities, all at the same time. The most challenging aspect is usually to estimate the number of components, as mixture models are non-nested so a log-likelihood test cannot be applied. Other methods are used, such as AIC, BIC, cross-validation and Bayesian methods (Chapter 5). Once the number of components is found, the unknown parameters are estimated by maximum likelihood or Bayesian methods.

4.11   The normal linear model

The most widely applied statistical model is the normal linear model. A lin-ear model is defined essentially by two main hypotheses. Given the explanatory variables $X_1, \ldots, X_p$ and the response variable Y, with $x_{1i}, \ldots, x_{pi}$ the observed values of the explanatory variables $X_1, \ldots, X_p$ corresponding to the ith observa-tion, the first hypothesis supposes that the corresponding observations $Y_1, \ldots, Y_n$ of the response variable Y are independent random variables, each normally distributed with different expected values $\mu_1, \ldots, \mu_n$ and equal variance σ_2:

$$E(Y_i | X_{1i} = x_{1i}, \ldots, X_{pi} = x_{pi}) = \mu_i,$$

$$\text{Var}(Y_i | X_{1i} = x_{1i}, \ldots, X_{pi} = x_{pi}) = \sigma^2, \qquad i = 1, \ldots, n.$$

Dropping the conditioning terms in the expressions for the means and variances, we can write:

$$E(Y_i) = \mu_i,$$

$$\text{Var}(Y_i) = \sigma^2, \qquad i = 1, \ldots, n.$$

For each $i = 1, \ldots, n$, let

$$\mathbf{x}_{i\bullet} = \begin{bmatrix} x_{0i} \\ x_{1i} \\ \cdots \\ x_{pi} \end{bmatrix} = \begin{bmatrix} 1 \\ x_{1i} \\ \cdots \\ x_{pi} \end{bmatrix}, \boldsymbol{\beta} = \begin{bmatrix} \beta_0 \\ \beta_1 \\ \cdots \\ \beta_p \end{bmatrix}.$$

The second hypothesis states that the mean value of the response variable is a linear combination of the explanatory variables:

$$\mu_i = \mathbf{x}'_{i\bullet}\boldsymbol{\beta} = \beta_0 + \beta_1 x_{1i} + \cdots + \beta_p x_{pi}, i = 1, \ldots, n.$$

In matrix terms, setting $\mathbf{Y} = (Y_1, \ldots, Y_n)'$,

$$\mathbf{X} = \begin{bmatrix} \mathbf{x}'_{1\bullet} \\ \vdots \\ \mathbf{x}'_{n\bullet} \end{bmatrix} = \begin{bmatrix} x_{01} & x_{11} & \cdots & x_{p1} \\ x_{02} & x_{12} & \cdots & x_{p2} \\ \cdots & \cdots & \cdots & \cdots \\ x_{0n} & x_{1n} & \cdots & x_{pn} \end{bmatrix}$$

and with $\boldsymbol{\beta}$ as above, the previous two hypotheses can be summarised by saying that $\mathbf{Y}$ is a multivariate normal variable with mean vector $E(\mathbf{Y}) = \boldsymbol{\mu} = \mathbf{X}\boldsymbol{\beta}$ and variance–covariance matrix $\boldsymbol{\Sigma} = E[(\mathbf{Y} - \boldsymbol{\mu})(\mathbf{Y} - \boldsymbol{\mu})'] = \sigma^2\mathbf{I}_n$, where $\mathbf{I}_n$ is the identity matrix of order n.

4.11.1 Main inferential results

Under the previous assumptions, we can derive some important inferential results that build on the theory in Section 4.3.

Result 1. From a point estimate point of view, it can be demonstrated that the least squares fitted parameters in Section 4.3 coincide with the maximum likelihood estimators of $\boldsymbol{\beta}$. We will use $\hat{\boldsymbol{\beta}}$ to refer to either of the two estimators.

Result 2. A confidence interval for a slope coefficient of the regression plane is

$$\beta = \hat{\beta} \pm t_{n-p-1}\left(1 - \frac{\alpha}{2}\right)\text{se}\left(\hat{\beta}\right),$$

where $t_{n-p-1}(1 - \alpha/2)$ is the $100(1 - \alpha/2)$ percentile of a Student's t distribution with $n - p - 1$ degrees of freedom and se $\left(\hat{\beta}\right)$ is an estimate of the standard error of $\hat{\beta}$.

Result 3. To test the hypothesis that a slope coefficient is 0, a rejection region is given by

$$R = \left\{ |T| \geq t_{n-p-1}\left(1 - \frac{\alpha}{2}\right) \right\}, \text{ where } T = \frac{\hat{\beta}}{\text{se}\left(\hat{\beta}\right)}.$$

If the observed absolute value of the statistic T is contained in the rejection region, the null hypothesis of slope equal to 0 is rejected, and the slope coefficient is *statistically significant*. In other words, the explanatory variable considered significantly influences the response variable. Conversely, when the observed absolute value of the statistic T falls outside the rejection region, the explanatory variable is not significant. Alternatively, it is possible to calculate the p-value of the test, the probability of observing a value of T greater in absolute value than the observed value. If this p-value is small (e.g. lower than $\alpha = 0.05$), this means that the observed value is very distant from the null hypothesis, therefore the null hypothesis is rejected (i.e. the slope coefficient is significant).

Result 4. To test whether a certain regression plane, with p explanatory variables, constitutes a significant linear model, it can be compared with a trivial model, with only the intercept. The trivial model, set to be the null hypothesis H_0, is obtained by simultaneously setting all slope coefficients to 0. The regression plane will be significant when the null hypothesis is rejected. A rejection region is given by the inequality

$$F = \frac{R^2/p}{(1 - R^2)/(n - p - 1)} \geq F_{p,n-p-1}(1 - \alpha),$$

where R^2 is the coefficient of determination seen in Section 4.3 and $F_{p,n-p-1}(1 - \alpha)$ is the $100\,(1 - \alpha)$ percentile of an F distribution with p and $n - p - 1$ degrees of freedom. The degrees of freedom of the denominator represent the difference in dimension between the observation space (n) and the fitting plane ($p + 1$); those of the numerator represent the difference in dimension between the fitting plane ($p + 1$) and a fitting point (1) defined by the only intercept. A p-value for the test can be calculated, giving further support to the significance of the model.

Notice how we have introduced a precise threshold for evaluating whether a certain regression model is valid in making predictions, in comparison with the simple arithmetic mean. But this is a relative statement, which gives little indication of how well the linear model fits the data at hand. A statistic like this can be applied to cluster analysis, assuming that the available observations come from a normal distribution. Then the degrees of freedom are $c - 1$ and $n - c$. The statistic is called a pseudo-F statistic because, in the general case of a non-normal distribution for the observations, the statistic does not have an F distribution.

Result 5. To compare two nested regression planes that differ in a single explanatory variable, say the $(p + 1)$th, present in one model but not in the other, the simpler model can be set as null hypothesis H_0, so that the more complex model is chosen if the null hypotheses is rejected, and vice versa. A rejection region can be defined by the inequality

$$F = \frac{r^2_{Y,X_{p+1}|X_1,\dots,X_p}/1}{(1 - r^2_{Y,X_{p+1}|X_1,\dots,X_p})/n - p - 2} \geq F_{1,n-p-2}(1 - \alpha),$$

where $r^2_{Y,X_{p+1}|X_1,\dots,X_p}$ is the partial correlation coefficient between X_{p+1} and the response variable Y, conditional on all present explanatory variables, and $F_{1,n-p-2}(1 - \alpha)$ is the $100(1 - \alpha)$ percentile of an F distribution, with 1 and $n - p - 2$ degrees of freedom.

Notice that the degrees of freedom of the denominator represent the difference in dimension between the observation space (n) and the more complex fitting plane $(p + 2)$; the degrees of freedom of the numerator represent the difference in dimension between the more complex fitting plane $(p + 2)$ and the simpler one $(p + 1)$. Alternatively, we can do the comparison by calculating the p-value of the test. This can usually be derived from the output table that contains the decomposition of the variance, also called the analysis of variance (ANOVA) table. By substituting the definition of the partial correlation coefficient $r^2_{Y,X_{p+1}|X_1,\dots,X_p}$, we can write the test statistic as

$$F = \frac{\text{Var}(\hat{Y}_{p+1}) - \text{Var}(\hat{Y}_p)}{(\text{Var}(Y) - \text{Var}(\hat{Y}_{p+1}))/(n - p - 2)},$$

therefore this F test statistic can be interpreted as the ratio between the additional variance explained by the $(p + 1)$th variable and the mean residual variance. In other words, it expresses the relative importance of the $(p + 1)$th variable. This test is the basis of a process which chooses the best model from a collection of possible linear models that differ in their explanatory variables. The final model is chosen through a series of hypothesis tests, each comparing two alternative models. The simpler of the two models is taken as the null hypothesis and the more complex model as the alternative hypothesis.

As the model space will typically contain many alternative models, we need to choose a search strategy that will lead to a specific series of pairwise comparisons. There are at least three alternative approaches. The forward selection procedure starts with the simplest model, without explanatory variables. It then complicates it by specifying in the alternative hypothesis H_1 a model with one explanatory variable. This variable is chosen to give the greatest increase in the explained variability of the response. The F test is used to verify whether or not the added variable leads to a significant improvement with respect to the model in H_0. In the negative case the procedure stops and the chosen model is the model in H_0 (i.e. the simplest model). In the affirmative case the model in H_0 is rejected and replaced with the model in H_1. An additional explanatory variable (chosen as before) is then inserted in a new model in H_1, and a new comparison is made. The procedure continues until the F test does not reject the model in H_0, which thus becomes the final model.

The backward elimination procedure starts with the most complex model, containing all the explanatory variables. It simplifies it by making the null hypotheses H_0 equal to the original model minus one explanatory variable. The eliminated variable is chosen to produce the smallest decrease in the explained variability of the response. The F test is used to verify whether or not the elimination of this variable leads to a significant improvement with respect to the model in H_1. In the negative case the chosen model is the model in H_1 (i.e. the most complex model) and the procedure stops. In the affirmative case the complex model in H_1 is rejected and replaced with the model in H_0. An additional variable is dropped (chosen as before) and the resulting model is set as H_0, then a new comparison is made. The procedure continues until the F test rejects the null hypothesis. Then the chosen model is the model in H_1.

The stepwise procedure is essentially a combination of the previous two. It begins with no variables; variables are then added one by one according to the forward procedure. At each step of the procedure, a backward elimination is carried out to verify whether any of the added variables should be removed.

Whichever procedure is adopted, the final model should be the same. This is true most of the time but it cannot be guaranteed. The significance level used in the comparisons is an important parameter as the procedure is carried out automatically by the software and the software uses the same level for all comparisons. For large samples, stepwise procedures are often rather unstable in finding the best models. It is not a good idea to rely solely on stepwise procedures for selecting models.

4.12 Generalised linear models

For several decades, the linear model has been the main statistical model for data analysis. However, in many situations the hypothesis of linearity is not realistic. The second restrictive element of the normal linear model is the assumption of normality and constant variance of the response variable. In many applications the observations are not normally distributed nor do they have a constant variance, and this limits the usefulness of the normal linear model. Developments in statistical theory and computing power during the 1960s allowed researchers to take their techniques for linear models and develop them in other contexts. It turns out that many of the 'nice' proprieties of the normal distribution are shared by a wider class of statistical models known as the exponential family of distributions.

The numerical calculations for the parameter estimates have also benefited from refinements; as well as working on linear combinations $\mathbf{X}\beta$, we can now work on functions of linear combinations such as $g(\mathbf{X}\beta)$. Improved computer hardware and software have helped with effective implementation, culminating in generalised linear models (Nelder and Wedderburn, 1972). In the normal linear model, the base distribution is the normal distribution; in generalised linear models it is one of the exponential family of distributions.

4.12.1   The exponential family

Consider a single random variable Y whose density function (or discrete probability function) depends on a single parameter θ (possibly vector-valued). The probability distribution of the variable is said to belong to the exponential family if the density can be written in the form

$$f(y; \theta) = s(y) t(\theta) e^{a(y)b(\theta)},$$

where a, b, s and t are known functions.

The symmetry existing between y and θ becomes more evident if the previous equation is rewritten in the form

$$f(y; \theta) = \exp\left[a(y) b(\theta) + c(\theta) + d(y)\right],$$

where $s(y) = \exp[d(y)]$ and $t(\theta) = \exp[c(\theta)]$. If $a(y) = y$, the previous distribution is said to be in canonical form, and $b(\theta)$ is called the natural parameter of the distribution. If there are other parameters (say, ϕ), besides the parameter of interest θ, they are considered as nuisance parameters that are usually dealt with as if they were known. Many familiar distributions belong to the exponential family; here are some of them.

Poisson distribution

The Poisson distribution is usually used to model the probability of observing integer numbers, corresponding to counts in a fixed period of time (e.g. the number of customers entering a supermarket in an hour, or the number of phone calls received in a call centre in a day). The Poisson distribution is a discrete distribution that assigns a non-zero probability to all the non-negative integers. It is parameterised by a parameter that represents the mean value of the counts. If a random variable Y has a Poisson distribution with mean λ, its discrete probability function is

$$f(y; \lambda) = \frac{\lambda^y e^{-\lambda}}{y!},$$

where y takes the values $0, 1, 2, \ldots$. Through simple algebra it is possible to rewrite the density as

$$f(y; \lambda) = \exp\left[y\log\lambda - \lambda - \log y!\right],$$

which shows that the Poisson distribution belongs to the exponential family, in canonical form, with natural parameter $b(\theta) = \log\lambda$.

Normal distribution

The normal distribution is a continuous distribution that assigns a positive density to each real number. If Y is a normal random variable with mean μ and variance σ^2, its density function is

$$f(y; \mu) = \frac{1}{(2\pi\sigma^2)^{1/2}}\exp\left[-\frac{1}{2\sigma^2}(y - \mu)^2\right],$$

where μ is usually the parameter of interest and σ^2 is considered a nuisance parameter. The density can be rewritten as

$$f(y; \mu) = f(y; \mu) = \exp\left[y\frac{\mu}{\sigma^2} - \frac{\mu^2}{2\sigma^2} - \frac{1}{2}\log(2\pi\sigma^2) - \frac{y^2}{2\sigma^2}\right],$$

which shows that the normal distribution belongs to the exponential family, in canonical form, with natural parameter $b(\theta) = \mu/\sigma^2$.

Binomial distribution

The binomial distribution is used to model the probability of observing a number of 'successes' (or events of interest) in a series of n independent binary trials (e.g. how many among the n customers of a certain supermarket buy a certain product, or how many among n loans assigned to a certain credit institution have a good end). The binomial distribution is a discrete distribution that assigns a non-zero probability to all the non-negative integers between 0 and n, representing the completed trials. It is parameterised by n and by the parameter π, which represents the probability of obtaining a success in each trial. Suppose that the random variable Y represents the number of successes in n binary independent experiments, in which the probability of success is always equal to π. Then Y has a binomial distribution with discrete probability function

$$f(y; \pi) = \binom{n}{y} \pi^y (1 - \pi)^{n-y},$$

where y takes the values $0, 1, 2, \ldots, n$. This function can be rewritten as

$$f(y; \pi) = \exp\left[y\log\left(\frac{\pi}{1 - \pi}\right) + n\log(1 - \pi) + \log\binom{n}{y}\right],$$

which shows that the binomial distribution belongs to the exponential family, in canonical form, with natural parameter $b(\theta) = \log[\pi/(1 - \pi)]$.

The exponential family of distribution is a very general class that contains these three important probability models. The advantage of the general form is that it is possible to obtain inferential results common to all the distributions belonging to it. We will not dwell on these results, limiting ourselves to describing some important consequences for data analysis. For more details, see Agresti (1990) or Dobson (2002).

4.12.2 Definition of generalised linear models

A generalised linear model takes a function of the mean value of the response variable and relates it to the explanatory variables through an equation having linear form. It is specified by three components: a random component, which identifies the response variable Y and assumes a probability distribution for it; a systematic component, which specifies the explanatory variables used as predictors in the model; and a link function, which describes the functional relation between the systematic component and the mean value of the random component.

Random component

For a sample of size n, the random component of a generalised linear model is described by the sample random variables $Y_1, \ldots, Y_n$; these are independent, each has a distribution in exponential family form that depends on a single parameter θ_i, and each is described by the density function

$$f(y_i; \theta_i) = \exp\left[y_i b(\theta_i) + c(\theta_i) + d(y_i)\right].$$

All the distributions for the Y_i have to be of the same form (e.g. all normal or all binomial) but their θ_i parameters do not have to be the same.

Systematic component

The systematic component specifies the explanatory variables and their roles in the model. It is given by a linear combination

$$\eta = \beta_1 x_1 + \cdots + \beta_p x_p = \sum_{j=1}^{p} \beta_j x_j.$$

The linear combination η is called the linear predictor. The X_j represent the covariates, whose values are known (e.g. they can derive from the data matrix). The β_j are the parameters that describe the effect of each explanatory variable on the response variable. The values of the parameters are generally unknown and have to be estimated from the data. The systematic part can be written in the following form:

$$\eta_i = \sum_{j=1}^{p} \beta_j x_{ij}, \qquad i = 1, \ldots, n,$$

where x_{ij} is the value of the jth explanatory variable for the ith observation. In matrix form, we have

$$\eta = X\beta,$$

where η is a vector of order $n \times 1$, X is a matrix of order $n \times p$, called the model matrix, and β is a vector of order $p \times 1$, called the parameter vector.

Link function

The third component of a generalised linear model specifies the link between the random component and the systematic component. Let the mean value of Y_i be denoted by

$$\mu_i = E(Y_i), i = 1, \ldots, n.$$

The link function specifies which function of μ_i linearly depends on the explanatory variables through the systematic component η_i. Let $g(\mu_i)$ be a (monotone and differentiable) function of μ_i. The link function is defined by

$$g(\mu_i) = \eta_i = \sum_{j=1}^{p} \beta_j x_{ij}, i = 1, \ldots, n.$$

Table 4.5   Main canonical links.

Distribution	Canonical link
Normal	$g(\mu_i) = \mu_i$
Binomial	$g(\mu_i) = \log\left(\frac{\pi_i}{1-\pi_i}\right)$
Poisson	$g(\mu_i) = \log \mu_i$

In other words, the link function describes how the explanatory variables affect the mean value of the response variable, that is, through the (not necessarily linear) function g. How do we chose g? In practice, the more commonly used link functions are canonical and define the natural parameter of the particular distribution as a function of the mean response value. Table 4.5 shows the canonical links for the three important distributions in Section 4.12.1. The same table can be used to derive the most important examples of generalised linear models. The simplest link function is the normal one. It directly models the mean value through the identity link $\eta_i = \mu_i$, thereby specifying a linear relationship between the mean response value and the explanatory variables:

$$\mu_i = \beta_0 + \beta_1 x_{i1} + \cdots + \beta_p x_{ip}.$$

The normal distribution and the identity link give rise to the normal linear model for continuous response variables (Section 4.11).

The binomial link function models the logarithm of the odds as a linear function of the explanatory variables:

$$\log\left(\frac{\mu_i}{1-\mu_i}\right) = \beta_0 + \beta_1 x_{i1} + \cdots + \beta_p x_{ip}, i = 1, \ldots, n.$$

This type of link, called the logit link, is appropriate for binary response variables, as in the binomial model. A generalised linear model that uses the binomial distribution and the logit link is a logistic regression model (Section 4.4). For a binary response variable, econometricians often use the probit link, which is not canonical. This assumes that

$$\Phi^{-1}(\mu_i) = \beta_0 + \beta_1 x_{i1} + \cdots + \beta_p x_{ip}, i = 1, \ldots, n,$$

where Φ^{-1} is the inverse of the cumulative normal distribution function.

The Poisson canonical link function specifies a linear relationship between the logarithm of the mean response value and the explanatory variables:

$$\log(\mu_i) = \beta_0 + \beta_1 x_{i1} + \cdots + \beta_k x_{ik}, i = 1, \ldots, n.$$

A generalised linear model that uses the Poisson distribution and a logarithmic link is a log-linear model; it constitutes the main data mining tool for describing associations between the available variables (see Section 4.13).

Inferential results

We now consider inferential results that hold for the whole class of generalised linear models; we will apply them logistic regression and log-linear models. Parameter estimates are usually obtained using the method of maximum likelihood. The method computes the derivative of the log-likelihood with respect to each coefficient in the parameter vector β and sets it equal to zero, similarly to the linear regression context in Section 4.3. But unlike what happens with the normal linear model, the resultant system of equations is non-linear in the parameters and does not generally have a closed-form solution. So to obtain maximum likelihood estimators of β we need to use iterative numerical methods, such as the Newton–Raphson method or Fisher's scoring method; for more details, see Agresti (1990) or Hand *et al.* (2001).

Once the parameter vector β is estimated, its significance is usually assessed by hypothesis testing. We will now see how to verify the significance of each parameter in the model. Later we will compute the overall significance of a model in the context of model comparison. Consider testing the null hypothesis $H_0 : \beta_i = 0$ against the alternative $H_1 : \beta_i \neq 0$. A rejection region for H_0 can be defined using the asymptotic procedure known as Wald's test. If the sample size is sufficiently large, the statistic

$$Z = \frac{\hat{\beta}_i}{\sigma(\hat{\beta}_i)}$$

is approximately distributed as standardised normal; here $\sigma(\hat{\beta}_i)$ denotes the standard error of the estimator in the numerator, Therefore, to decide whether to accept or reject the null hypothesis, we can construct the rejection region

$$R = \{|Z| \geq z_{1-\alpha/2}\},$$

where $z_{1-\alpha/2}$ is the $100(1 - \alpha/2)$ percentile of the standard normal distribution. Alternatively, we can find the p-value and see whether it is less than a predefined significance level (e.g. $\alpha = 0.05$). If $p < \alpha$ then H_0 is rejected. The square of the Wald statistic Z has a chi-squared distribution with 1 degree of freedom, for large samples. That means it is legitimate for us to construct a rejection region or to assess a p-value.

Rao's score statistic, often used as an alternative to the Wald statistic, computes the derivative of the observed log-likelihood function evaluated at the parameter values set by the null hypothesis, $\beta_i = 0$. Since the derivative is zero at the point of maximum likelihood, the absolute value of the score statistic tends to increase as the maximum likelihood estimate $\hat{\beta}_i$ gets more distant from zero. The score statistic is equal to the square of the ratio between the derivative and its standard error, and it is also asymptotically distributed as chi-squared with 1 degree of freedom. For more details on hypotheses testing using generalised linear models, see Dobson (2002), McCullagh and Nelder (1989) or Azzalini (1992).

Model comparison

Fitting a model to data can be interpreted as a way of replacing a set of observed data values with a set of estimated values obtained during the fitting process. The number of parameters in the model is generally much lower than the number of observations in the data. We can use these estimated values to predict future values of the response variable from future values of the explanatory variables. In general the fitted values, say $\hat{\mu}_i$, will not be exactly equal to the observed values, y_i. The problem is to establish the distance between the $\hat{\mu}_i$ and the y_i. In Chapter 5 we will start from this simple concept of distance between observed values and fitted values, then show how it is possible to construct statistical measures to compare statistical models and, more generally, data mining methods. In this section we will consider the deviance and the Pearson statistic, two measures for comparing the goodness of fit of different generalised linear models.

The first step in evaluating a model's goodness of fit is to compare it with the models that produce the best fit and the worst fit. The best-fit model is called the saturated model; it has as many parameters as observations (n) and therefore leads to a perfect fit. The worst-fit model is called the null model; it has only one intercept parameter and leaves all response variability unexplained. The saturated model attributes the whole response variability to the systematic component. In practice the null model is too simple and the saturated model is not informative because it does completely reproduces the observations. However, the saturated model is a useful comparison when measuring the goodness of fit of a model with p parameters. The resulting quantity is called the deviance, and it is defined as follows for a model M (with p parameters) in the class of the generalised linear models:

$$G^2(M) = -2\log\left\{\frac{L\left(\hat{\beta}(M)\right)}{L\left(\hat{\beta}(M^*)\right)}\right\},$$

where the quantity in the numerator is the likelihood function, calculated using the maximum likelihood parameter estimates under model M, denoted by $\hat{\beta}(M)$; and the quantity in the denominator is the likelihood function of the observations, calculated using the maximum likelihood parameter estimates under the saturated model M^*. The expression in curly brackets is called the likelihood ratio, and it seems an intuitive way to compare two models in terms of the likelihood they receive from the observed data. Multiplying the natural logarithm of the likelihood ratio by -2, we obtain the maximum likelihood ratio test statistic.

The asymptotic distribution of the G^2 statistic (under H_0) is known: for a large sample size, $G^2(M)$ is approximately distributed as chi-squared with $n - k$ degrees of freedom, where n is the number of observations and k is the number of the estimated parameters under model M, corresponding to the number of explanatory variables plus one (the intercept). The logic behind the use of G^2 is as follows. If the model M that is being considered is good, then the value of its maximised likelihood will be closer to that of the maximised likelihood under the saturated model M^*. Therefore 'small' values of G^2 indicate a good fit.

The asymptotic distribution of G^2 can provide a threshold beneath which to declare the simplified model M as a valid model. Alternatively, the significance can be evaluated through the p-value associated with G^2. In practice the p-value represents the area to the right of the observed value for G^2 in the χ^2_{n-k} distribution. A model M is considered valid when the observed p-value is large (e.g. greater than 5%). The value of G^2 alone is not sufficient to judge a model, because G^2 increases as more parameters are introduced, similarly to what happens to R^2 in the regression model. However, since the threshold value generally decreases with the number of parameters, by comparing G^2 and the threshold we can reach a compromise between goodness of fit and model parsimony.

The overall significance of a model can also be evaluated by comparing it against the null model. This can be done by taking the difference in deviance between the model considered and the null model to obtain the statistic

$$D = -2 \log \left\{ \frac{L\left(\hat{\beta}(M_0)\right)}{L\left(\hat{\beta}(M)\right)} \right\}.$$

Under the null hypothesis that the null model is true, D is asymptotically distributed as χ^2_p, where p is the number of explanatory variables in model M. This can be obtained by noting that $D = G_2(M_0) - G_2(M)$ and recalling that the two deviances independent and asymptotically distributed as chi-squared random variables. From the additive property of the chi-squared distribution, it follows that the degrees of freedom of D are $(n-1) - (n-p-1) = p$. The model considered is accepted (i.e. the null model in the null hypothesis is rejected) if the p-value is small. This is equivalent to the difference D between the log-likelihoods being large. Rejection of the null hypothesis implies that at least one parameter in the systematic component is significantly different from zero. Statistical software often gives the log-likelihood maximised for each model of analysis, corresponding to $-2 \log(L(\hat{\beta}(M)))$, which can be seen as a score of the model under consideration. To obtain the deviance, this should be normalised by subtracting $-2 \log(L(\hat{\beta}(M^*)))$. On the other hand, D is often given with the corresponding p-values. This is analogous to the statistic in Result 4 in Section 4.11.1 (page 114).

More generally, following the same logic used in the derivation of D, any two models can be compared in terms of their deviances. If the two models are nested (i.e. the systematic component of one of them is obtained by eliminating some terms from the other one), the difference between the deviances is asymptotically distributed as chi-squared with $p - q$ degrees of freedom, where $p - q$ represents the number of variables excluded in the simpler model (which has q parameters) but not in the other one (which has p parameters). If the difference between the two is large (with respect to the critical value), the simpler model will be rejected in favour of the more complex model, and similarly when the p-value is small.

For the whole class of generalised linear models, it is possible to employ a formal procedure in searching for the best model. As with linear models, this procedure is usually forward, backward or stepwise elimination.

When the data analysed are categorical, or discretised to be such, an alternative to G^2 is Pearson's X^2:

$$X^2 = \sum_i \frac{(o_i - e_i)^2}{e_i}$$

where, for each category i, o_i represents the observed frequencies and e_i represents the frequencies expected according to the model under examination. As with the deviance G^2, we are comparing the fitted model (which corresponds to the e_i) and the saturated model (which corresponds to the o_i). However, the distance function is not based on the likelihood, but on direct comparison between observed and fitted values for each category. Notice that this statistic generalises the Pearson X^2 distance measure in Section 3.4. There the fitted model particularised to the model under which the two categorical variables were independent.

The Pearson statistic is asymptotically equivalent to G^2, therefore under H_0, $X^2 \approx \chi^2_{n-k}$. The choice between G^2 and X^2 depends on the goodness of the chi-squared approximation. In general, it can be said that X^2 is less affected by small frequencies, particularly when they occur in large data sets – data sets having many variables. The advantage of G^2 lies in its additivity, so it easily generalises to any pairwise model comparisons, allowing us to adopt a model selection strategy.

The statistics G^2 and X^2 indicate a model's overall goodness of fit; we need to do further diagnostic analysis to look for local lack of fit. Before fitting a model, it may be extremely useful to try some graphical representations. For example, we could plot the observed frequencies of the various categories, or functions of them, against the explanatory variables. It is possible to draw dispersion diagrams and fit straight lines for the response variable transformation described by the canonical link (e.g. the logit function for logistic regression). This can be useful in verifying whether the hypotheses behind the generalised linear model are satisfied. If not, the graph itself may suggest further transformations of the response variable or the explanatory variables. Once a model is chosen as the best and fitted to the data, our main diagnostic tool is to analyse the residuals. Unlike what happens in the normal linear model, for generalised linear models there are different definitions of the residuals. Here we consider the deviance residuals that are often used in applications. For each observation, the residual from the deviance is defined by the quantity

$$_D r_i = (y_i - \hat{\mu}_i) \sqrt{d_i}.$$

This quantity increases (or decreases) depending on the difference between the observed and fitted values of the response variable $(y_i - \hat{\mu}_i)$ and is such that $\sum_D r_i^2 = G^2$. In a good model, the deviance residuals should be randomly distributed around zero, so plot the deviance residuals against the fitted values. For a good model, the points in the plane should show no evident trend.

4.12.3 The logistic regression model

The logistic regression model is an important model, and we can use our general results to derive inferential results for it. The deviance of a model M takes the form

$$G^2(M) = 2 \sum_{i=1}^{n} \left[y_i \log \left(\frac{y_i}{n_i \hat{\pi}_i} \right) + (n_i - y_i) \log \left(\frac{n_i - y_i}{n_i - n_i \hat{\pi}_i} \right) \right],$$

where the $\hat{\pi}_i$ are the fitted probabilities of success, calculated on the basis of the estimated β parameters for model M. Observe the similarity between this form and

$$G^2 = 2 \sum_i o_i \log \frac{o_i}{e_i},$$

where o_i stands for the observed frequencies y_i and $n_i - y_i$, and e_i stands for the corresponding fitted frequencies $n_i \hat{\pi}_i$ and $n_i - n_i \hat{\pi}_i$. Note that G^2 can be interpreted as a distance function, expressed in terms of entropy differences between the fitted model and the saturated model.

The Pearson statistic for the logistic regression model, based on the X^2 distance, takes the form

$$Z^2 = \sum_{i=1}^{n} \frac{(y_i - n_i \hat{\pi}_i)^2}{n_i \hat{\pi}_i (1 - \hat{\pi}_i)}.$$

Both G^2 and X^2 can be used to compare models in terms of distances between observed and fitted values. The advantage of G^2 lies in its modularity. For instance, in the case of two nested logistic regression models M_A, with q parameters, and M_B, with p parameters ($q < p$), the difference between the deviances is given by

$$D = G^2(M_A) - G^2(M_B) = 2 \sum_{i=1}^{n} y_i \log \left(\frac{n_i \hat{\pi}_i^B}{n_i \hat{\pi}_i^A} \right) + (n_i - y_i) \log \left(\frac{n_i \hat{\pi}_i^B}{n_i \hat{\pi}_i^A} \right)$$

$$= 2 \sum_{i=1}^{n} o_i \log \left(\frac{e_i^B}{e_i^A} \right) \approx \chi^2_{p-q},$$

where $\hat{\pi}_i^A$ and $\hat{\pi}_i^B$ denote the success probability fitted, on the basis of models M_A and M_B, respectively. Note that the expression for the deviance boils down into an entropy measure between probability models, exactly as before. This is a general fact. The deviance residuals are defined by Finally, we remark that, in the logistic regression model,:

$$_D r_i = \pm (y_i - \hat{\pi}_i) 2^{1/2} \left[y_i \log \left(\frac{y_i}{n_i \hat{\pi}_i} \right) + (n_i - y_i) \log \left(\frac{n_i - y_i}{n_i - n_i \hat{\pi}_i} \right) \right]^{1/2}.$$

4.13   Log-linear models

We can distinguish symmetric and asymmetric generalised linear models. If the objective of the analysis is descriptive – to describe the associative structure among the variables – the model is called symmetric. If the variables are divided into two groups, response and explanatory – to predict the responses on the basis of the explanatory variables – the model is asymmetric. Asymmetric models we have seen are the normal linear model and the logistic regression model. We will now consider the best-known symmetric model, the log-linear model. The log-linear model is typically used for analysing categorical data, organised in contingency tables. It represents an alternative way to express a joint probability distribution for the cells of a contingency table. Instead of listing all the cell probabilities, this distribution can be described using a more parsimonious expression given by the systematic component.

4.13.1   Construction of a log-linear model

We now show how a log-linear model can be constructed, starting from three different distributional assumptions about the absolute frequencies of a contingency table, corresponding to different sampling schemes for the data in the table. For simplicity, but without loss of generality, we consider a two-way contingency table of dimension $I \times J$ (I rows and J columns).

Scheme 1
The cell counts are independent random variables that have a Poisson distribution. All the marginal counts, including the total number of observations n, are also random and Poisson distributed. As the natural parameter of a Poisson distribution with parameter m_{ij} is $\log(m_{ij})$, the relationship that links the expected value of each cell frequency m_{ij} to the systematic component is

$$\log\left(m_{ij}\right) = \eta_{ij},$$

for $i = 1, \ldots, I$ and $j = 1, \ldots, J$. In the linear and logistic regression models, the total amount of information (which determines the degrees of freedom) is described by the number of observations of the response variable (denoted by n), but in the log-linear model this corresponds to the number of cells of the contingency table. In the estimation procedure, the expected frequencies will be replaced by the observed frequencies, and this will lead us to estimate the parameters of the systematic component. For an $I \times J$ table there are two variables in the systematic component. Let the levels of the two variables be denoted by x_i and x_j, for $i = 1, \ldots, I$ and $j = 1, \ldots, J$. The systematic component can therefore be written as

$$\eta_i = u + \sum_i u_i x_i + \sum_j u_j x_j \sum_{ij} u_{ij} x_i x_j$$

This expression is called the log-linear expansion of the expected frequencies. The terms u_i and u_j describe the single effects of each variable, corresponding to the mean expected frequencies for each of their levels. The term u_{ij} describes the joint effect of the two variables on the expected frequencies. The term u is a constant that corresponds to the mean expected frequency over all table cells.

Scheme 2

The total number of observations n is not random, but a fixed constant. This implies that the relative frequencies follow a multinomial distribution. Such a distribution generalises the binomial to the case where there are more than two alternative events for the variable considered. The expected values of the absolute frequencies for each cell are given by $m_{ij} = n\pi_{ij}$. With n fixed, specifying a statistical model for the probabilities π_{ij} is equivalent to modelling the expected frequencies m_{ij}, as in Scheme 1.

Scheme 3

The marginal row (or column) frequencies are known. In this case it can be shown that the cell counts are distributed as a product of multinomial distributions. It is possible to show that we can define a log-linear model in the same way as before.

Properties of the log-linear model

Besides being parsimonious, the log-linear model allows us easily to incorporate in the probability distribution constraints that specify independence relationships between variables. For example, using results introduced in Section 3.4, when two categorical variables are independent, the joint probability of each cell probability factorises as $\pi_{ij} = \pi_{i+}\pi_{+j}$, for $i = 1, \ldots, I$ and $j = 1, \ldots, J$. And the additive property of the logarithms implies that

$$\log(m_{ij}) = \log n + \log \pi_{i+} + \log \pi_{+j}.$$

This describes a log-linear model of independence that is more parsimonious than the previous one, called the saturated model as it contains as many parameters as there are table cells. In general, to achieve a unique estimate on the basis of the observations, the number of terms in the log-linear expansion cannot be greater than the number of cells in the contingency table. This implies some constraints on the parameters of a log-linear model. Known as identifiability constraints, they can be defined in different ways, but we will use a system of constraints that equates to zero all the u-terms that contain at least one index equal to the first level of a variable. This implies that, for a 2×2 table, relative to the binary variables A and B, with levels 0 and 1, the most complex possible log-linear model (saturated) is defined by

$$\log\left(m_{ij}\right) = u + u_i^A + u_j^B + u_{ij}^{AB}$$

with constraints such that: $u_i^A \neq 0$ for $i = 1$ (i.e. if $A = 1$); $u_j^B \neq 0$ for $j = 1$ (i.e. if $B = 1$); and $u_{ij}^{AB} \neq 0$ for $i = 1$ and $j = 1$ (i.e. if $A = 1$ and $B = 1$). The notation reveals that, in order to model the four cell frequencies in the table, there is a constant term, u; two main effects terms that exclusively depend on a variable, u_i^A and u_j^B; and an interaction term that describes the association between the two variables, u_{ij}^{AB}. Therefore, following the stated constraints, the model establishes that the logarithms of the four expected cell frequencies are given by

$$\log(m_{00}) = u,$$

$$\log(m_{10}) = u + u_i^A,$$

$$\log(m_{00}) = u + u_j^B,$$

$$\log(m_{00}) = u + u_i^A + u_j^B + u_{ij}^{AB}.$$

4.13.2   Interpretation of a log-linear model

Logistic regression models with categorical explanatory variables (also called logit models) can be considered a particular case of the log-linear models. To clarify this point, consider a contingency table with three dimensions for variables A, B, C, and numbers of levels I, J, 2 respectively. Assume that C is the response variable of the logit model. A logit model is expressed by

$$\log\left(\frac{m_{ij1}}{m_{ij0}}\right) = \alpha + \beta_i^A + \beta_j^B + \beta_{ij}^{AB}.$$

All the explanatory variables of a logit model are categorical, so the effect of each variable (e.g. variable A) is indicated only by the coefficient (e.g. β_i^A) rather than by the product (e.g. βA). Besides that, the logit model has been expressed in terms of the expected frequencies, rather than probabilities, as in the last section. This is only a notational change, obtained through multiplying the numerator and the denominator by n. This expression is useful to show that the logit model which has C as response variable is obtained as the difference between the log-linear expansions of $\log(m_{ij1})$ and $\log(m_{ij0})$. Indeed, the log-linear expansion for a contingency table with three dimensions $I \times J \times 2$ has the more general form

$$\log\left(m_{ijk}\right) = u + u_i^A + u_j^B + u_k^C + u_{ij}^{AB} + u_{ik}^{AC} + u_{jk}^{BC} + u_{ijk}^{ABC}.$$

Substituting and taking the difference between the logarithms of the expected frequencies for $C = 1$ and $C = 0$:

$$\log\left(m_{ij1}\right) - \log\left(m_{ij0}\right) = u_1^C + u_{i1}^{AC} + u_{j1}^{BC} + u_{ij1}^{ABC}.$$

In other words, the u-terms that do not depend on the variable C cancel out. All the remaining terms depend on C. By eliminating the symbol C from the superscript, the value 1 from the subscript and relabelling the u-terms using α and

β, we arrive at the desired expression for the logit model. Therefore a logit model can be obtained from a log-linear model. The difference is that a log-linear model contains not only the terms that describe the association between the explanatory variables and the response – here the pairs *AC, BC* – but also the terms that describe the association between the explanatory variables – here the pair *AB*. Logit models do not model the association between the explanatory variables.

We now consider the relationship between log-linear models and odds ratios. The logarithm of the odds ratio between two variables is equal to the sum of the interaction u-terms that contain both variables. It follows that if in the log-linear expansion considered there are no u-terms containing both the variables A and B, say, then we obtain $\theta_{AB} = 1$; that is, the two variables are independent.

To illustrate this concept, consider a 2×2 table and the odds ratio between the binary variables A and B:

$$\theta = \frac{\theta_1}{\theta_2} = \frac{\pi_{1|1}/\pi_{0|1}}{\pi_{1|0}/\pi_{0|0}} = \frac{\pi_{11}/\pi_{01}}{\pi_{10}/\pi_{00}} = \frac{\pi_{11}\pi_{00}}{\pi_{01}\pi_{10}}.$$

Multiplying numerator and denominator by n^2 and taking logarithms:

$$\log(\theta) = \log(m_{11}) + \log(m_{00}) - \log(m_{10}) - \log(m_{01}).$$

Substituting for each probability the corresponding log-linear expansion, we obtain $\log(\theta) = u_{11}^{AB}$. Therefore the odds ratio between the variables A and B is $\theta = \exp(u_{11}^{AB})$. These previous relations, which are very useful for data interpretation, depend on the identifiability constraints we have adopted.

We have shown the relationship between the odds ratio and the parameters of a log-linear model for a 2×2 contingency table. This result is valid for contingency tables of higher dimension, provided the variables are binary and, as usually happens in a descriptive context, the log-linear expansion does not contain interaction terms between more than two variables.

4.13.3 Graphical log-linear models

A key instrument in understanding log-linear models, and graphical models in general, is the concept of conditional independence for a set of random variables; this extends the notion of statistical independence between two variables, seen in Section 3.4. Consider three random variables X, Y, and Z. X and Y are conditionally independent given Z if the joint probability distribution of X and Y, conditional on Z, can be decomposed into the product of two factors: the conditional density of X given Z and the conditional density of Y given Z. In formal terms, X and Y are conditionally independent of Z if $f(x, y|Z = z) = f(x|Z = z)f(y|Z = z)$ and we write $X \perp Y|Z$. An alternative way of expressing this concept is that the conditional distribution of Y on both X and Z does not depend on X. So, for example, if X is a binary variable and Z is a discrete variable, then for every z and y we have

$$f(y|X = 1, Z = z) = f(y|X = 0, Z = z) = f(y|Z = z).$$

The notion of (marginal) independence between two random variables (Section 3.4) can be obtained as a special case of conditional independence. As seen for marginal independence, conditional independence can simplify the expression for and the interpretation of log-linear models. In particular, it can be extremely useful in visualising the associative structure among all variables at hand, using the so-called independence graphs. Indeed, a subset of log-linear models, called graphical log-linear models, can be completely characterised in terms of conditional independence relationships and therefore graphs. For these models, each graph corresponds to a set of conditional independence constraints and each of these constraints can correspond to a particular log-linear expansion.

The study of the relationship between conditional independence statements, represented in graphs, and log-linear models has its origins in the work of Darroch *et al.* (1980). We explain this relationship through an example. For a systematic treatment, see Edwards (2000), Whittaker (1990) or Lauritzen (1996). We believe that the introduction of graphical log-linear models helps to explain the problem of model choice for log-linear models. Consider a contingency table of three dimensions, each corresponding to a binary variable so that the total number of cells in the contingency table is $2^3 = 8$. The simplest log-linear graphical model for a three-way contingency table assumes that the logarithm of the expected frequency of every cell is

$$\log\left(m_{jkl}\right) = u + u_j^A + u_k^B + u_l^C.$$

This model does not contain interaction terms between variables, therefore the three variables are mutually independent. In fact, the model can be expressed in terms of cell probabilities as $p_{jkl} = p_{j++}p_{+k+}p_{++l}$, where the symbol + indicates that the joint probabilities have been summed with respect to all the values of the relative index. Note that, for this model, the three odds ratios between the variables – (A, B), (A, C), (B, C) – are all equal to 1. To uniquely identify the model it is possible to use a list of the terms, called generators, that correspond to the maximal terms of interaction in the model. These terms are called maximals in the sense that their presence implies the presence of interaction terms between subsets of their variables. At the same time, their existence in the model is not implied by any other term. For the previous model of mutual independence, the generators are (A, B, C); they are the main effect terms as there are no other terms in the model. To graphically represent conditional independence statements, we can use conditional independence graphs. These are constructed by associating a node with each variable and by placing a link (technically, an edge) to connect a pair of variables whenever the corresponding random variables are dependent. For the cases of mutual independence we have described, there are no edges and therefore we obtain the representation in Figure 4.11.

Consider now a more complex log-linear model for the three variables, described by the log-linear expansion

$$\log\left(m_{jkl}\right) = u + u_j^A + u_k^B + u_l^C + u_{jk}^{AB} + u_{jl}^{AC}.$$

In this case, since the maximal terms of interaction are u_{jk}^{AB} and u_{jl}^{AC}, the generators of the model will be (AB, AC). Notice that the model can be reformulated

Figure 4.11 Conditional independence graph for the mutual independence case.

in terms of cell probabilities as

$$\pi_{jkl} = \frac{\pi_{jk+}\pi_{j+l}}{\pi_{j++}}$$

or, equivalently, as

$$\frac{\pi_{jkl}}{\pi_{j++}} = \frac{\pi_{jk+}}{\pi_{j++}} \frac{\pi_{j+l}}{\pi_{j++}}$$

which, in terms of conditional independence, states that

$$P\,(B=k, C=l\,|A=j) = P\,(B=k\,|A=j)\,P\,(C=l\,|A=j)\,.$$

The indicates that, in the conditional distribution (on A), B and C are independent – in symbols, $B \perp C | A$. Therefore, the conditional independence graph of the model is as in Figure 4.12. It can be demonstrated that, in this case, the odds ratios between all variable pairs are different from 1, while the two odds ratios for the two-way table between B and C, conditional to A, are both equal to 1.

We finally consider the most complex (saturated) log-linear model for the three variables,

$$\log\left(m_{jkl}\right) = u + u_j^A + u_k^B + u_l^C + u_{jk}^{AB} + u_{jl}^{AC} + u_{kl}^{BC} + u_{jkl}^{ABC},$$

which has (ABC) as generator. This model does not establish any conditional independence constraints on cell probabilities. Correspondingly, all odds ratios, marginal and conditional, will be different from 1. The corresponding conditional independence graph will be completely connected. The previous model (AB, AC) can be considered as a particular case of the saturated model, obtained by setting $u_{kl}^{BC} = 0$ for all k and l and $u_{jkl}^{ABC} = 0$ for all j, k, l. Equivalently, it is obtained by removing the edge between from B and C in the completely connected graph, which corresponds to imposing the constraint that B and C are independent conditionally on A. Notice that the mutual independence model is a particular

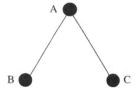

Figure 4.12 Conditional independence graph corresponding to $B \perp C | A$.

case of the saturated model obtained setting $u_{kl}^{BC} = u_{jl}^{AC} = u_{jk}^{AB} = u_{jkl}^{ABC} = 0$, for all j, k, l, or by removing all three edges in the complete graph. Consequently, the differences between log-linear models can be expressed in terms of differences between the parameters or as differences between graphical structures. We think it is easier to interpret differences between graphical structures.

All the models in this example are graphical log-linear models. In general, graphical log-linear models are definable as log-linear models that have as generators the cliques of the conditional independence graph. A clique is a subset of completely connected and maximal nodes in a graph. For example, in Figure 4.12 the subsets AB and AC are cliques, and they are the generators of the model. On the other hand, the subsets formed by the isolated nodes A, B and C are not cliques. To better understand the concept of a graphical log-linear model, consider a non-graphical model for the trivariate case. Take the model described by the generator (AB, AC, BC):

$$\log(m_{jkl}) = u + u_j^A + u_k^B + u_l^C + u_{jk}^{AB} + u_{jl}^{AC} + u_{kl}^{BC}.$$

Although this model differs from the saturated model by the absence of the three-way interaction term u_{jkl}^{ABC}, its conditional independence graph is the same, with one single clique, ABC. Therefore, since the model generator is different from the set of cliques, the model is not graphical. To conclude, in this section we have obtained a remarkable equivalence relation between: conditional independence statements, graphical representations and probability models, with the probability models represented in terms of cell probabilities, log-linear models or sets of odds ratios.

4.13.4 Log-linear model comparison

For log-linear models, including graphical log-linear models, we can apply the inferential theory derived for generalised linear models. We now turn to model comparison, because the use of conditional independence graphs permits us to interpret model comparison and choice between log-linear models in terms of comparisons between sets of conditional independence constraints. In data mining problems the number of log-linear models to compare increases rapidly with the number of variables considered. Therefore a valid approach may be to restrict the class of models. In particular, a parsimonious and efficient way to analyse large contingency tables is to consider interaction terms in the log-linear expansion that involve at most two variables. The log-linear models in the resulting class are all graphical. Therefore we obtain an equivalence relationship between the absence of an edge between two nodes, say i and j, conditional independence between the corresponding variables, X_i and X_j (given the remaining ones), and nullity of the interaction parameter indexed by both of them.

As we saw with generalised linear models, the most important tool for comparing models is the deviance. All three sampling schemes for log-linear models lead to an equivalent expression for the deviance. Consider, for simplicity, a log-linear model to analyse three categorical variables. The deviance of a model

M is

$$G^2(M) = 2 \sum_{jkl} n_{jkl} \log \left(\frac{n_{jkl}}{\hat{m}_{jkl}} \right) = 2 \sum o_i \log \frac{o_i}{e_i},$$

where $\hat{m}_{jkl} = np_{jkl}$, the p_{jkl} are the maximum likelihood estimates of the cell probabilities, the o_i are the observed cell frequencies and the e_i are the cell frequencies estimated according to the model M. Notice the similarity with the deviance expression for the logistic regression model. What changes is essentially the way in which the cell probabilities are estimated. In the general case of a p-dimensional table, the definition is the same, but the index set changes:

$$G^2(M_0) = 2 \sum_{i \in I} n_i \log \left(\frac{n_i}{\hat{m}_i^0} \right),$$

where, for a cell i belonging to the index set I, n_i is the frequency of observations in the ith cell and $\hat{m}_i^0$ are the expected frequencies for the considered model M_0. For model comparison, two nested models M_0 and M_1 can be compared using the difference between their deviances:

$$D = G_0^2 - G_1^2 = 2 \sum_{i \in I} n_i \log \left(\frac{n_i}{\hat{m}_i^0} \right) - 2 \sum_{i \in I} n_i \log \left(\frac{n_i}{\hat{m}_i^1} \right) = 2 \sum_{i \in I} n_i \log \left(\frac{\hat{m}_i^1}{\hat{m}_i^0} \right).$$

As in the general case, under H_0, D has an asymptotic chi-squared distribution whose degrees of freedom are obtained taking the difference in the number of parameters for models M_0 and M_1.

The search for the best log-linear model can be carried out using a forward, backward or stepwise procedure. For graphical log-linear models we can also try adding or removing edges between variables rather than adding or removing interaction parameters. In the backward procedure we compare the deviance between models that differ by the presence of an edge and at each step we eliminate the less significant edge; the procedure stops when no arc removals produce a p-value greater than the chosen significance level (e.g. 0.05). In the forward procedure we add the most significance edges one at time until no arc additions produce a p-value lower than the chosen significance level.

4.14 Graphical models

Graphical models are models that can be specified directly through conditional independence relationships among the variables, represented in a graph. Although the use of graphics with statistical models is not a new idea, the work of Darroch *et al.* (1980) has combined the two concepts in a new and illuminating way. They showed that a subset of the log-linear models, the log-linear graphical models, can be easily interpreted in terms of conditional independence relationships. This finding has led to the development of a wide class of statistical models, known as graphical models, which provide considerable flexibility in specifying models to analyse databases of whatever dimension, containing both qualitative and quantitative variables, and admitting both symmetric and asymmetric relationships. Graphical models contain, as special cases, important classes of generalised linear models, such as the three seen in this book. For a detailed treatment, see Lauritzen (1996), Whittaker (1990) or Edwards (2000).

Here are some definitions we will need. A graph $G = (V, E)$ is a structure consisting of a finite number V of vertices (nodes) that correspond to the variables present in the model, and a finite number of edges between them. In general, the causal influence of one variable on another is indicated by a directed edge (shown using an arrow), while a symmetric association is represented by an undirected edge (shown using a line). Figure 4.13 is an example of an undirected graph, containing only undirected edges. Figure 4.14 is a directed graph for the same type of application, where we have introduced a new variable, X, which corresponds to the return on investments of the enterprises. We have made a distinction between vertices that represent categorical variables (empty circles) and vertices that represent continuous variables (filled circles).

Two vertices X and Y belonging to V are adjacent, written $X \sim Y$, if they are connected by an undirected arc; that is, if both the pairs (X, Y) and (Y, X) belong to E. A node X is a parent of the node Y, written $X \rightarrow Y$, if they are connected by a directed edge from X to Y; that is, if $(X, Y) \in E$ and $(Y, X) \notin E$.

A complete graph is a graph in which all pairs of vertices are connected by an edge. A sequence of vertices $X_0, \ldots, X_n$ such that $X_{i-1} \sim X_i$ for $i = 1, \ldots, n$ is called a path of length n. A graph is connected when there exists at least one path between each pair of vertices.

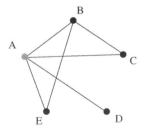

Figure 4.13   Conditional independence graph for the final selected model.

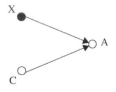

Figure 4.14   Example of a directed graph.

These are only some of the properties we can define for graphical models, but they are sufficient to understand the probabilistic assumptions implied by a conditional independence graph.

In general, a graphical model is a family of probability distributions that incorporates the rules of conditional independence described by a graph. The key to interpreting a graphical model is the relationship between the graph and the probability distribution of the variables. It is possible to distinguish three types of graph, related to the three main classes of graphical models:

- *Undirected graphs* are used to model symmetric relations among the variables (Figure 4.13); they give rise to the symmetric graphical models.
- *Directed graphs* are used to model asymmetric relations among the variables (Figure 4.14); they give rise to recursive graphical models, also known as probabilistic expert systems.
- *Chain graphs* contain both undirected and directed edges, therefore they can model both symmetric and asymmetric relationships; they give rise to graphical chain models (Cox and Wermuth, 1996).

4.14.1 Symmetric graphical models

In symmetric graphical models, the probability distribution is Markovian with respect to the specified undirected graph. This is equivalent to imposing on the distribution a number of probabilistic constraints known as Markov properties. The constraints can be expressed in terms of conditional independence relationships. Here are two Markov properties and how to interpret them:

- For the *pairwise Markov property*, if two nodes are not adjacent in the fixed graph, the two corresponding random variables will be conditionally independent, given the others. On the other, hand, if the specified probability distribution is such that $X \sim Y|$ others, the edge between the nodes corresponding to X and Y has to be omitted from the graph.
- For the *global Markov property*, if two sets of variables, U and V, are graphically separated by a third set of variables, W, then $U \sim V|W$. For example, consider four discrete random variables, W, X, Y, and Z, whose conditional independence relations are described by the graph in Figure 4.15, from which we have that W and Z are separated from X and Y, and Y and Z are separated from X. A Markovian distribution with respect to the graph in Figure 4.15 has to satisfy the global Markov property and therefore we have that $W \sim Z|(X, Y)$ and $Y \sim Z|(W, X)$.

It is useful to distinguish three types of symmetric graphical models:

- Discrete graphical models coincide with log-linear graphical models and are used when all the available variables are categorical.
- Graphical Gaussian models are used when the joint distribution of all variables is multivariate Gaussian.
- Mixed graphical models are used for a mixture of categorical variables and multivariate Gaussian variables.

STATISTICAL DATA MINING

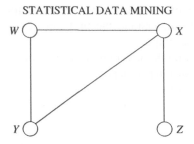

Figure 4.15   Illustration of the global Markov property.

We have seen discrete graphical models in the Section 4.13.3. A similar type of symmetric model, useful for descriptive data mining, can be introduced for continuous variables. An exhaustive description of these models can be found in Whittaker (1990), who has called them Gaussian graphical models even though they were previously known in the statistical literature as covariance selection models (Dempster, 1972). For these models, it is assumed that $Y = (Y_1, \ldots, Y_q)$ is a vector of continuous variables with a normal multivariate distribution. Markovian properties allow us to show that two variables are conditionally independent on all the others, if and only if the element corresponding to the two variables in the inverse of the variance–covariance matrix is null. This is equivalent to saying that the partial correlation coefficient between the two variables, given the others, is null. In terms of conditional independence graphs, given four variables X, Y, W, Z, if the elements of the inverse of the variance–covariance matrix $k_{x,z}$ and $k_{y,w}$ were null, the edges between the nodes X and Z and the nodes Y and W would have to be absent. From a statistical viewpoint, a graphical Gaussian model and, equivalently, a graphical representation are selected by successively testing hypotheses of edge removal or addition. This is equivalent to testing whether the corresponding partial correlation coefficients are zero.

Notice how the treatment of the continuous case is similar to the discrete case. This has allowed us to introduce a very general class of mixed symmetric graphical models. We now introduce them in a rather general way, including continuous and discrete graphical models as special cases. Let $V = \Gamma \cup \Delta$ be the vertex set of a graph, partitioned into a set of $|\Gamma|$ continuous variables, and a set of $|\Delta|$ discrete variables. If with each vertex v is associated a random variable X_v, the whole graph is associated with a random vector $X_V = (X_v, v \sim V)$. A mixed graphical model is defined by a conditional Gaussian distribution for the vector X_V. Partition X_V into a vector X_Δ containing the categorical variables, and a vector X_Γ containing the continuous variables. Then X_V follows a conditional Gaussian distribution if it satisfies the following

- $p(i) = P(X_\Delta = i) > 0;$
- $p(X_\Gamma | X_\Delta = i) = N_{|\Gamma|}(\xi(i), \Sigma(i)),$

where N denotes a Gaussian distribution of dimension $|\Gamma|$, with mean vector $\xi(i) = K(i)^{-1}h(i)$ and variance–covariance matrix $\Sigma(i) = K(i)^{-1}$, positive definite. In words, a random vector has a conditional Gaussian distribution if the distribution of the categorical variables is described by a set of positive cell probabilities (this could happen through the specification of a log-linear model) and the continuous variables are distributed, conditional on each joint level of the categorical variables, as a Gaussian distribution with a null mean vector and a variance–covariance matrix that can, in general, depend on the levels of the categorical variables.

From a probabilistic viewpoint, a symmetric graphical model is specified by a graph and a family of probability distributions, which has Markov properties with respect to it. However, to use graphical models in real applications, it is necessary to completely specify the probability distribution, usually by estimating the unknown parameters on the basis of the data. This inferential task, usually accomplished by maximum likelihood estimation, is called quantitative learning. Furthermore, in data mining problems it is difficult to avoid uncertainty when specifying a graphical structure, so alternative graphical representations have to be compared and selected, again on the basis of the available data; this constitutes the so-called structural learning task, usually tackled by deviance-based statistical tests.

To demonstrate this approach, we can return to the European software industry application in Section 4.6 and try to describe the associative structure among all seven random variables considered. The graph in Figure 4.16 is based on hypotheses formulated through subject matter research by industrial economics experts; it shows conditional independence relationships between the available variables. One objective of the analysis is to verify whether the graph in Figure 4.16 can be simplified, maintaining a good fit to the data (structural learning). Another objective is to verify some research hypothesis on the sign of the association between some variables (quantitative learning).

We begin by assuming a probability distribution of conditional Gaussian type and, given the reduced sample size (51 observations), a homogeneous model (Lauritzen, 1996). A homogeneous model means that we assume the variance of the continuous variable does not depend on the level of the qualitative variables. So that we can measure explicitly the effect of the continuous variable Y on the

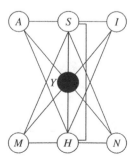

Figure 4.16   Initial conditional independence graph.

qualitative variables, we have decided to maintain, in all models considered, a link between Y and the qualitative variables, even when it is not significant on the basis of the data. It is opportune to start the selection from the given initial research graph (Figure 4.16). Since the conditional Gaussian distribution has to be Markovian with respect to the previous graph, all the parameters containing the pairs $\{M, A\}, \{N, I\}, \{M, N\}, \{M, I\}, \{A, N\}, \{A, I\}$ have to be 0, hence the total number of parameters in the model corresponding to Figure 4.17 is 29. Considering the small number of available observations, this model is clearly overparameterised.

A very important characteristic of graphical models is to permit local calculations on each clique of the graph (Frydenberg and Lauritzen, 1989). For instance, as the above model can be decomposed in four cliques, it is possible to estimate the parameters separately, on each clique, using the 51 available observations to estimate the 17 parameters of each marginal model. In fact, on the basis of a backward selection procedure using a significance level of 5%, Giudici and Carota (1992) obtained the final structural model shown in Figure 4.17. From the figure we deduce that the only direct significant associations between qualitative variables are between the pairs $\{H, I\}, \{N, S\}$ and $\{N, H\}$. These associations depend on the revenue Y but not on the remaining residual variables. Concerning quantitative learning, the same authors have used their final model to calculate the odds ratios between the qualitative variables, conditionally on the level of Y. They obtained the following estimated conditional odds, relative to the pairs $\{H, I\}, \{N, S\}$ and $\{N, H\}$:

$$\hat{\Theta}_{IH|R} = \exp(0.278 + 0.139R), \quad \text{therefore } \hat{\Theta}_{IH|R} > 1 \text{ for } R > 0.135$$

(all enterprises);

$$\hat{\Theta}_{NH|R} = \exp(-2.829 + 0.356R), \quad \text{therefore } \hat{\Theta}_{NH|R} > 1 \text{ for } R > 2856$$

(only one enterprise);

$$\hat{\Theta}_{NS|R} = \exp(-0.827 - 0.263R), \quad \text{therefore } \hat{\Theta}_{NS|R} > 1 \text{ for } R < 23.21$$

(for 23 enterprises).

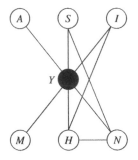

Figure 4.17   Final conditional independence graph.

The signs of the association can be summarised as follows: the association between I and H is positive; the association between N and H is substantially negative; the association between N and S is positive only for enterprises having revenues less than the median.

From an economic viewpoint, these associations have a simple interpretation. The relationship between I and H confirms that enterprises that adopt a strategy of incremental innovations tend to increase their contacts with enterprises in the hardware sector. The strategy of creating radically new products is based on an opposite view. Looking at contacts exclusively within the software sector, small enterprises (having revenues less than the median) tend to fear their innovations could be stolen or imitated and they tend not to make contacts with other small companies. Large companies (having revenues greater than the median) do not fear imitation and tend to increase their contacts with other companies.

4.14.2 Recursive graphical models

Directed or recursive graphical models, also known as probabilistic expert systems, are an important and sophisticated tool for predictive data mining. Their fundamental assumption is that the variables can be partially ordered so that every variable is logically preceded by a set of others. This precedence can be interpreted as a probabilistic dependency and, more strongly, as a causal dependency. Both interpretations exist in the field of probabilistic expert systems and this is reflected in the terminology: casual network if there is a causal interpretation, belief network if there is no causal interpretation.

To specify any recursive model, we need to specify a directed graph that establishes the (causal) relationships among the variables. Once this graph is specified, a recursive model is obtained by using a probability distribution that is Markov with respect to the graph (e.g. Lauritzen, 1996). The Markov properties include the following factorisation property of the probability distribution:

$$f(x_1, \ldots, x_p) = \prod_{i=1}^{p} f(x_i|\mathrm{pa}(x_i)),$$

where $\mathrm{pa}(x_i)$ indicates the parent nodes of each of the p nodes considered. This specifies that the probability distribution of the p variables is factorised into a series of local terms, each of which describes the dependency of each of the considered variables, x_i, on the set of relevant explanatory variables, $\mathrm{pa}(x_i)$. It is a constructive way to specify a directed graphical model using a (recursive) sequence of asymmetric models, each of which describes the predictor set of each variable.

The conditional independence graphs are therefore directed, because the edges represent ordered pairs of vertices. They are also constrained to be acyclic: no sequence of connected vertices has to form a loop. Figure 4.18 is an example of an acyclic directed conditional independence graph. It states, for instance, that $V_3 \perp V_2 | V_4$, and it corresponds to a recursive model that can be specified, for

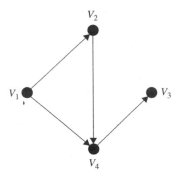

Figure 4.18 Example of a directed acyclic graph.

instance, by the factorisation

$$p(V_1, V_2, V_3, V_4) = p(V_1)\,p(V_2|V_1)\,p(V_4|V_1, V_2)\,p(V_3|V_4).$$

When it comes to specifying the local probability distributions (e.g. the distributions of V_1, V_2, V_3, V_4), we could specify a recursive model as a recursion of generalised linear models. For example, Figure 4.18 corresponds to a model defined by a linear regression of V_3 on the explanatory variable V_4, of V_4 on the explanatory variables V_1 and V_2, and of V_2 on the explanatory variable V_1.

From a predictive data mining perspective, the advantage of recursive models is that they are simple to specify, interpret and summarise, thanks to the factorisation of the model into many local models of reduced dimensions. On the other hand, they do involve more sophisticated statistical thinking, especially in the context of structural learning. However, few mainstream statistical packages implement these types of model. Directed graphical models are typically used in artificial intelligence applications, where they are known as probabilistic expert systems or Bayesian networks (e.g. Heckerman, 1997).

Another important special case of directed graphical models are Markov chain models, which are particularly useful for modelling time series data. In particular, first-order Markov models are characterised by a short memory, and assume that $X_{i+1} \perp X_{j<i}\,|\{X_i\}$, where X_{i+1} is a random variable that describe the value of a certain quantity at time $i+1$, i is the previous time point and $j < i$ denote time points further back in time. In other words, the future occurrence of a random variable X_{i+1} does not depend on the past values $X_{j<i}$, if the present value X_i is known. If the data can be modelled by a Markov chain model, the joint probability distribution will factorise as

$$f(x_1, \ldots, x_p) = \prod_{i=1}^{p} f(x_i|x_{i-1}).$$

Recursive graphical models can also be used to classify observations, in which case they are more commonly known as Bayesian networks. One of the simplest and most useful Bayesian networka is the naive Bayes model. It arises when there is one qualitative response variable that can assume M values, corresponding to

M classes. The goal is to classify each observation into one of the three classes, and p explanatory variables can be used, described by a random vector $\mathbf{X}$. As we have seen, the Bayes classifier allocates each observation to the class C_i that maximises the posterior probability

$$p(C_i|\mathbf{X} = \mathbf{x}) = p(\mathbf{X} = \mathbf{x}|C_i)p(C_i)/p(\mathbf{X} = \mathbf{x}).$$

The naive Bayes model corresponds to a special case of this rule, obtained when the explanatory variables that appear in the vector $\mathbf{X}$ are assumed to be conditionally independent given the class label. A Bayesian network can be seen as a more sophisticated and more realistic version of a Bayes classifier, which establishes that among the explanatory variables there are relationships of conditional dependence specified by a directed graph.

4.14.3 Graphical models and neural networks

We have seen that the construction of a statistical model is a long and conceptually complex process that requires the formulation of a series of formal hypotheses. On the other hand, a statistical model allows us to make predictions and simulate scenarios on the basis of explicit rules that are easily scalable – rules that can be generalised to different data. We have seen how computationally intensive techniques require a lighter analytical structure, allowing us to find precious information rapidly from large volumes of data. Their disadvantages are low transparency and low scalability. Here is a brief comparison to help underline the different concepts. We shall compare neural networks and graphical models; they can be seen as rather general examples of computational methods and statistical methods, respectively.

The nodes of a graphical model represent random variables, whereas in neural networks they are computational units, not necessarily random. In a graphical model an edge represents a probabilistic conditional dependency between the corresponding pair of random variables, whereas in a neural network an edge describes a functional relation between the corresponding nodes. Graphical models are usually constructed in three phases: (a) the qualitative phase establishes the conditional independence relationships among the random variables; (b) the probabilistic phase associates the graph with a vector of random variables having a Markovian distribution with respect to the graph; (c) the quantitative phase assigns the parameters (if known) that characterise the distribution in (b). Neural networks are constructed in three similar phases: (a) the qualitative phase establishes the organisation of the layers and the relationships among them; (b) the functional phase specifies the functional relationships between the layers; (c) the quantitative phase fixes the weights (if known) associated with the connections among the different nodes.

We believe that these two methodologies can be used in a complementary way. Taking a graphical model and introducing latent variables – variables that are not observed – confers two extra advantages. First, it allows us to represent a multilayer perceptron as a graphical model, so we can take formal statistical methods valid for graphical models and use them on neural networks (e.g. confidence intervals, rejection regions, deviance comparisons). Second, the use of a neural network in a preliminary phase could help to reduce the structural complexity of graphical models, reducing the number of variables and edges present, and doing it in a more computationally efficient way. Adding latent variables to graphical models, corresponding to purely computational units, allows us to enrich the model with non-linear components, as occurs with neural networks. For more on the role of latent variables in graphical models, see Cox and Wermuth (1996).

4.15   Survival analysis models

Survival analysis (e.g. Singer and Willet, 2003) focuses on the time between entry to a study and some subsequent event. The standard approaches to survival analysis are stochastic; that is, the times at which events occur are assumed to be realisations of random processes. It follows that T, the event time for some particular individual, is a random variable with a probability distribution. A useful, model-free or non-parametric approach for all random variables uses the cumulative distribution function (e.g. Hougaard 1995). The cumulative distribution function of a variable T, denoted by $F(t)$, tells us the probability that the variable will be less than or equal to some value t; that is, $F(t) = P\{T \leq t\}$. If we know the value of F for every value of t, then we know all there is to know about the distribution of T. In survival analysis it is more common to work with a closely related function called the survivor function, defined as

$$S(t) = P\{T > t\} = 1 - F(t).$$

If the event of interest is a death (or, equivalently, a churn), the survivor function gives the probability of surviving beyond t. Because S is a probability we know that it is bounded by 0 and 1; and because T cannot be negative, we know that $S(0) = 1$. Often the objective is to compare survivor functions for different subgroups in a sample (clusters, regions, etc.). If the survivor function for one group is always higher than the survivor function for another group, then the first group clearly lives longer than the second group.

For continuous variables, another common way of describing their probability distribution is the probability density function. This function is defined as

$$f(t) = \frac{dF(t)}{dt} = -\frac{dS(t)}{dt};$$

that is, the probability density function is just the derivative or slope of the cumulative distribution function. For continuous survival data, the hazard function is more popular than the probability density function as a way of describing distributions. The hazard function (e.g. Allison 1995) is defined as

$$h(t) = \lim_{\varepsilon t \to 0} \frac{P\{t \leq T < t + \varepsilon t | T \geq t\}}{\varepsilon t}.$$

This definition quantifies the instantaneous risk that an event will occur at time t. Since time is continuous, the probability that an event will occur exactly at time t is necessarily 0. But we can talk about the probability that an event occurs in the small interval between t and $t + \varepsilon t$ and we also want to make this probability conditional on the individual surviving to time t. For this formulation the hazard function is sometimes described as a conditional density and, when events are repeatable, the hazard function is often referred to as the intensity function. The survival function, the probability density function and the hazard function are equivalent ways of describing a continuous probability distribution. Another formula expresses the hazard in terms of the probability density function:

$$h(t) = \frac{f(t)}{S(t)},$$

which leads to

$$h(t) = -\frac{d}{dt} \log S(t).$$

Integrating both sides of this equation gives an expression for the survival function in terms of the hazard function:

$$S(t) = \exp\left(-\int_0^t h(u)\, du \right).$$

The hazard is a dimensional quantity that expresses the number of events per interval of time.

The first step in the analysis of survival data (for descriptive purposes) is to plot the survival function and the risk. The survival function is estimated by the Kaplan–Meier method (Kaplan and Meier, 1958). Suppose that there are K distinct event times, $t_1 < t_2 < \ldots < t_k$. At each time t_j there are n_j individuals who are said to be at risk of an event; that is, they have not experienced an event and they have not been censored prior to time t_j. If any cases are censored at exactly t_j, there are also considered to be at risk at t_j. Let d_j be the number of individuals who die at time t_j. The Kaplan–Meier estimator is defined as

$$\hat{S}(t) = \prod_{j:t_j \leq t} \left(1 - \frac{d_j}{n_j} \right), \quad t_1 \leq t \leq t_k,$$

This formula says that, for a given time t, we take all the event times that are less than or equal to t. For each of these event times, we compute the quantity in parentheses, which can be interpreted as the conditional probability of surviving to time t_{j+1}, given that one has survived to time t_j. Then we multiply all of these survival probability together.

For predictive purposes, the most popular model is Cox regression (Cox, 1972). Cox proposed a proportional hazards model and a new method of estimation that came to be called partial likelihood or, more accurately, maximum partial likelihood. We start with the basic model that does not include time-dependent covariate or non-proportional hazards. The model is usually written as

$$h(t_{ij}) = h_0(t_j) \exp\left[\beta_1 X_{1ij} + \beta_2 X_{2ij} + \cdots + \beta_p X_{pij}\right].$$

It says that the hazard for individual i at time t is the product of two factors: a baseline hazard function that is left unspecified, and a linear combination of a set of p fixed covariates, which is then exponentiated. The baseline function can be regarded as the hazard function for an individual whose covariates all have values 0. The model is called the proportional hazards model because the hazard for any individual is a fixed proportion of the hazard for any other individual. To see this, take the ratio of the hazards for two individuals i and j:

$$\frac{h_i(t)}{h_j(t)} = \exp\left\{\beta_1 \left(x_{i1} - x_{j1}\right) + \cdots + \beta_p \left(x_{ip} - x_{jp}\right)\right\}.$$

What is important about this equation is that the baseline cancels out of the numerator and denominator. As a result, the ratio of the hazards is constant over time.

4.16   Further reading

In the chapter we have reviewed the most important data mining methodologies, beginning in Sections 4.1–4.8 with those that do not strictly require a probabilistic model. The first section explained how to calculate a distance matrix from a data matrix. Sometimes we want to build a data matrix from a distance matrix, and one solution is the method of multidimensional scaling (e.g. Mardia *et al.*, 1979). Having applied multidimensional scaling, it is possible to represent the row vectors (statistical observations) and the column vectors (statistical variables) in a unique plot called a biplot; this helps us to make interesting interpretations of the scores obtained. In general, biplots are used with tools for reducing dimensionality, such as principal component analysis and correspondence analysis. For an introduction to this important theme, see Gower and Hand (1996); in a data mining context, see Hand *et al.* (2001).

The next section was concerned with cluster analysis, probably one of the best-known techniques used in the statistical analysis of multidimensional data. An interesting extension of cluster analysis is fuzzy classification; this allows a 'weighted' allocation of the observations to the clusters (Zadeh, 1977).

Multivariate linear regression is best dealt with using matrix notation. For an introduction to matrix algebra in statistics, see Searle (1982). The logistic regression model is for predicting categorical variables. The estimated category probabilities can then be used to classify statistical observations in groups, according to a supervised method. Probit models, well known in economics, are essentially the same as logistic regression models, once the logistic link is replaced by an inverse Gaussian link (e.g. Agresti, 1990).

Local model rules are still at an embryonic stage of development, at least from a statistical viewpoint. Association rules seem ripe for a full statistical treatment, and we have covered them in some depth. However, we have only briefly referred to retrieval-by-content methods, which are expected to gain in importance in the foreseeable future, especially with reference to text mining. See Hand *et al.* (2001) on retrieval by content and Zanasi (2003) on text mining. The statistical understanding of local models will be an important area for future research.

Tree models are probably the most popular data mining technique. A more detailed account can be found in advanced data mining textbooks, such as Hastie *et al.* (2001) and Hand *et al.* (2001). These texts offer a statistical treatment; a computational treatment can be found in Han and Kamber (2001). The original works on CART and CHAID are Breiman *et al.* (1984) and Kass (1980).

Neural networks and support vector machines are important classes of supervised models that originated in the machine learning communities. We have not considered support vector machines in this book because they do not provide explicit and transparent solutions and therefore are rarely used in business data mining problems. The literature on neural networks is vast; for a classical statistical approach, see Bishop (1995); for a Bayesian approach, consult Neal (1996). Support vector machines are discussed by 1998), Vapnik (1995.

In recent years statistical models of increasing complexity have been developed that closely resemble neural networks but have a more statistical structure. Examples are the projection pursuit models, the generalised additive models, and the multivariate adaptive regression spline models. For a review, see Cheng and Titterington (1994) and Hastie *et al.* (2001).

In Section 4.9–4.15 we have reviewed the main statistical models for data mining applications. Their common feature is the presence of probability modelling. We began with methods for modelling uncertainty and inference; there are many textbooks on this. One to consult is Mood *et al.* (1991); another is Azzalini (1992), which takes more of a modelling viewpoint. Non-parametric models are distribution free, as they do not require heavy preliminary assumptions. They may be very useful, especially in an exploratory context. For a review of non-parametric methods, see Gibbons and Chakraborti (1992). Semiparametric models, based on mixture models, can provide a powerful probabilistic approach to cluster analysis. For an introductory treatment from a data mining viewpoint, see Hastie *et al.* (2001).

Introduction of the Gaussian distribution allows us to bring regression methods into the field of normal linear models. For an introduction to the normal linear model, see Mood *et al.* (1991) or a classic econometrics text such as Greene (2000). The need for predictive tools for response variables that are neither continuous nor normal led to the development of generalised linear models from the normal linear model. For an introduction, see Dobson (2002), McCullagh and Nelder (1989), Nelder and Wedderburn (1972) and Agresti (1990).

Log-linear models are an important class generalised linear models. They are symmetric models and are mainly used to obtain the associative structure among categorical variables, whose observations are classified in multiple contingency tables. Graphical log-linear models are particularly useful for data interpretation. For an introduction to log-linear models, see Agresti (1990) or Christensen

(1997). For graphical log-linear models it is better to consult texts on graphical models, such as Whittaker (1990).

We introduced the concept of conditional independence (and dependence); graphical representation of conditional independence relationships allowed us to take what we saw for graphical log-linear models and generalise it to a wider class of statistical models, known as graphical models. Graphical models are very general statistical models for data mining. In particular, they can adapt to different analytical objectives, from predicting multivariate response variables (recursive models) to finding associative structure (symmetric models), in the presence of both qualitative and quantitative variables. For an introduction to graphical models, see Edwards (2000), Whittaker (1990) or Lauritzen (1996). For directed graphical models, also known as probabilistic expert systems, see Cowell *et al.* (1999) or Jensen (1996).

CHAPTER 5

Model evaluation

This chapter discusses methods for choosing among alternative models. In Chapter 4 we looked at the problem of comparing the various statistical models within the theory of statistical hypothesis testing. With this in mind we looked at the sequential procedures (forward, backward or stepwise) that allow us to choose a model by means of a sequence of pairwise comparisons. These criteria are generally not applicable to computational data mining models, which do not necessarily have an underlying probability model and therefore do not allow us to apply the statistical theory of hypothesis testing.

A particular data problem can often be tackled using several classes of models. For instance, in a problem in predictive classification it is possible to use logistic regression and tree models as well as neural networks.

Furthermore, model specification, hence model choice, is determined by the type of the variables. After exploratory analysis the data may be transformed or some observations may be eliminated; this will also affect the variables. So we need to compare models based on different sets of variables present at the start. For example, how do we take a linear model having the original explanatory variables and compare it with a model having principal components as explanatory variables?

All this suggests the need for a systematic study of how to compare and evaluate statistical models for data mining. In this chapter we will review the most important methods. As these criteria will be frequently used and compared in Part II of the text, this chapter will just offer a brief systematic summary without giving examples. We begin by introducing the concept of discrepancy for a statistical model; it will make us look further at comparison criteria based on statistical tests. Although this leads to a very rigorous methodology, it allows only a partial ordering of the models. Scoring functions are a less structured approach developed in the field of information theory. We explain how they give each model a score, which puts them into some kind of complete order. Another group of criteria has been developed in the machine learning field. We introduce the main computational criteria, such as cross-validation. These criteria have the advantage of being generally applicable but might be data-dependent and might require long computation times. We then introduce the very important concept of combining several models via model averaging, bagging and boosting. Finally, we introduce a group of criteria that are specifically tailored to business data

Applied Data Mining for Business and Industry, 2e   P. Giudici, S. Figini
© 2009 John Wiley & Sons, Ltd

mining. These criteria compare models in terms of the losses arising from their application. These criteria have the advantage of being easy to understand, but they still need formal improvements and mathematical refinements.

5.1   Criteria based on statistical tests

The choice of the statistical model used to describe a database is one of the main aspects of statistical analysis. A model is either a simplification or an approximation of reality and therefore it does not entirely reflect reality. As we have seen in Chapter 4, a statistical model can be specified by a discrete probability function or by a probability density function $f(x)$; this is what is considered to be 'underlying the data' or, in other words, it is the generating mechanism of the data. A statistical model is usually specified up to a set of unknown quantities that have to be estimated from the data at hand.

Often a density function is parametric or, rather, it is defined by a vector of parameters $\Theta = (\theta_1, \ldots, \theta_I)$ such that each value θ of Θ corresponds to a particular density function, $f_\theta(x)$. A model that has been correctly parameterised for a given unknown density function $f(x)$ is a model that gives $f(x)$ for particular values of the parameters. We can select the best model in a non-parametric context by choosing the distribution function that best approximates the unknown distribution function. But first of all we consider the notion of a distance between a model f, which is the 'true' generating mechanism of the data, and model g, which is an approximating model.

5.1.1   Distance between statistical models

We can use a distance function to compare two models, say g and f. As explained in Section 4.1, there are different types of distance function; here are the most important ones.

In the categorical case, a distance is usually defined by comparing the estimated discrete probability distributions, denoted by f and g. In the continuous case, we often refer to two variables, X_f and X_g, representing fitted observation values obtained with the two models.

Entropy distance
The entropy distance is used for categorical variables and is related to the concept of heterogeneity reduction (Section 3.4). It describes the proportional reduction of the heterogeneity between categorical variables, as measured by an appropriate index. Because of its additive property, the entropy is the most popular heterogeneity measure for this purpose. The entropy distance of a distribution g from a target distribution f is

$$_E d = \sum_i f_i \log \frac{f_i}{g_i},$$

which is the form of the uncertainty coefficient (Section 3.4), but also the form taken by the G^2 statistic. The G^2 statistic can be employed for most probabilistic data mining models. It can therefore be applied to prediction problems, such as logistic regression and directed graphical models, but also to descriptive problems, such as log-linear models and probabilistic cluster analysis. It also finds application with non-probabilistic models, such as classification trees. The Gini index can also be used as a measure of heterogeneity.

Chi-squared distance

The chi-squared distance between a distribution g and a target f is

$$\chi^2 d = \sum_i \frac{(f_i - g_i)^2}{g_i}$$

which corresponds to a generalisation of the Pearson's statistic seen in Section 3.4. This distance is used both for descriptive and predictive problems in the presence of categorical data, as an alternative to the entropy distance. It does not require an underlying probability model; we have seen its application within the CHAID decision trees algorithm.

0–1 distance

The 0–1 distance applies to categorical variables, and it is typically used for supervised classification problems. It is defined as

$$_{0-1}d = \sum_{r=1}^{n} 1(X_{fr} - X_{gr})$$

where $1(w - z) = 1$ if $w = z$ and 0 otherwise. It measures the distance in terms of a 0–1 function that counts the number of correct matches between the classifications carried out using the two models. Dividing by the number of observations give the misclassification rate, probably the most important evaluation tool in predictive classification models, such as logistic regression, classification trees and nearest-neighbour models.

Euclidean distance

Applied to quantitative variables, the Euclidean distance between a distribution g and a target f is

$$_2d(X_f, X_g) = \sqrt{\sum_{r=1}^{n} (X_{fr} - X_{gr})^2}.$$

It represents the distance between two vectors in the Cartesian plane. The Euclidean distance leads to the R^2 index and to the F test statistics. Furthermore, by squaring it and dividing by the number of observations we obtain the mean squared error. The Euclidean distance is widely used, especially for continuous predictive models, such as linear models, regression trees, and continuous probabilistic expert systems. But it is also used in descriptive models for the observations, such as cluster analysis and Kohonen maps. Notice that it does not necessarily require an underlying probability model. When there is an underlying probability model, it is usually the normal distribution.

Uniform distance

The uniform distance applies to comparisons between distribution functions. For two distribution functions F, G with values in [0, 1], the uniform distance is

$$\sup_{0 \le t \le 1} |F(t) - G(t)|.$$

The uniform distance is used in non-parametric statistics such as the Kolmogorov–Smirnov statistic (Section 4.10), which is typically employed to assess whether a non-parametric estimator is valid. But it is also used to verify whether a specific parametric model, such as the Gaussian model, is a good approximation to a non-parametric model.

5.1.2 Discrepancy of a statistical model

The distances in Section 5.1.1 can be used to define the notion of discrepancy for a model. Suppose that f represents an unknown density, and let $g = p_\theta$ be a family of density functions (indexed by a vector of parameters, θ). The discrepancy of a statistical model g, with respect to a target model f, can be defined using the Euclidean distance as

$$\Delta(f, p_\theta) = \sum_{i=1}^{n} (f(x_i) - p_\theta(x_i))^2.$$

For each observation, $i = 1, \ldots, n$, this discrepancy (which is a function of the parameters θ) considers the error made by replacing f with g.

If we knew f, the real model, we would be able to determine which of the approximating statistical models, different choices for g, would be the best, in the sense of minimising the discrepancy. Therefore, the discrepancy of g (due to the parametric approximation) can be obtained as the discrepancy between the unknown probability model and the best parametric statistical model, $p_{\theta_0}^{(I)}$:

$$\Delta(f, p_{\theta_0}^{(I)}) = \sum_{i=1}^{n} (f(x_i) - p_{\theta_0}^{(I)}(x_i))^2.$$

However, since f is unknown we cannot identify the best parametric statistical model. Therefore we will substitute f with a sample estimate denoted by $p_{\hat{\theta}}^{(I)}(x)$, for which the I parameters are estimated on the basis of the data. The discrepancy between this sample estimate of $f(x)$ and the best statistical model is called the discrepancy of g (due to the estimation process):

$$\Delta(p_{\hat{\theta}}^{(I)}, p_{\theta_0}^{(I)}) = \sum_{i=1}^{n} (p_{\hat{\theta}}^{(I)}(x_i) - p_{\theta_0}^{(I)}(x_i))^2.$$

Now we have a discrepancy that is a function of the observed sample. Bear in mind the complexity of g. To get closer to the unknown model, it is better to choose a family where the models have a large number of parameters. In other words, the discrepancy due to parametric approximation is smaller for more complex models. However, the sample estimates obtained with the more complex model tend to overfit the data, resulting in a greater discrepancy due to estimation. The aim in model selection is to find a compromise between these opposite effects of parametric approximation and estimation. The total discrepancy, known as the discrepancy between the function f and the sample estimate $p_{\hat{\theta}}^{(I)}$, takes both these factors into account. It is given by the equation

$$\Delta(f, p_{\hat{\theta}}^{(I)}) = \sum_{i=1}^{n} (f(x_i) - p_{\hat{\theta}}^{(I)}(x_i))^2$$

which represents the algebraic sum of two discrepancies, one from the parametric approximation and one from the estimation process. Generally, minimisation of the first discrepancy favours complex models, which are more adaptable to the data, whereas minimisation of the second discrepancy favours simple models, which are more stable.

The best statistical model to approximate f will be the model $p_{\hat{\theta}}^{(I)}$ that minimizes the total discrepancy. The total discrepancy can rarely be calculated in practice as the density function $f(x)$ is unknown. Therefore instead of minimizing the total discrepancy, the model selction problem is solved by minimizing the total expected discrepancy, $E\Delta(f, p_{\hat{\theta}}^{(I)})$, where the expectation is taken with respect to the sample probability distribution. Such an estimator defines an evaluation criterion, for a model with I parameters. Model choice will then be based on comparing the corresponding estimators, known as minimum discrepancy estimators.

5.1.3 Kullback–Leibler discrepancy

We now consider how to derive a model evaluation criterion. To define a general estimator we consider, rather than the Euclidean discrepancy we have already met, a more general discrepancy known as the Kullback–Leibler discrepancy (or divergence). The Kullback–Leibler discrepancy can be applied to observations of any type; it derives from the entropy distance and is given by

$$\Delta_{KL}(f, p_{\hat{\theta}}^{(I)}) = \sum_{i} f(x_i) \log \left(\frac{f(x_i)}{p_{\hat{\theta}}^{(I)}(x_i)} \right).$$

This can be easily mapped to the expression for the G^2 deviance; then the target density function corresponds to the saturated model. The best model can be interpreted as the one with a minimal loss of information from the true unknown distribution. Like the entropy distance, the Kullback–Leibler discrepancy is not symmetric.

We can now show that the statistical tests used for model comparison are based on estimators of the total Kullback–Leibler discrepancy. Let p_θ denote a probability density function parameterised by the vector $\Theta = (\theta_1, \ldots, \theta_I)$. The sample values $x_1, \ldots, x_n$ are a series of independent observations that are identically distributed, therefore the sample density function is expressed by the equation

$$L(\theta; x_1, \ldots, x_n) = \prod_{i=1}^{n} p_\theta(x_i)$$

Let $\hat{\theta}_n$ denote the maximum likelihood estimator of the parameters, and let the likelihood function L be calculated at this point. Taking the logarithm of the resulting expression and multiplying it by $-1/n$, we get

$$\Delta_{KL}(f, p_{\hat{\theta}}^{(I)}) = -\frac{1}{n} \sum_{i=1}^{n} \log[p_{\hat{\theta}}^{(I)}(x_i)],$$

known as the sample Kullback–Leibler discrepancy function. This expression can be shown to be the maximum likelihood estimator of the total expected Kullback–Leibler discrepancy of a model p_θ. Notice that the Kullback–Leibler discrepancy gives a score to each model, corresponding to the mean (negative) log-likelihood of the observations, which is equal to

$$2n\Delta_{KL}(f, p_{\hat{\theta}}^{(I)}) = -2 \sum_{i=1}^{n} \log[p_{\hat{\theta}}^{(I)}(x_i)].$$

The Kullback–Leibler discrepancy is fundamental to selection criteria developed in the field of statistical hypothesis testing. These criteria are based on a successive comparisons between pairs of alternative models. Let us suppose that the expected discrepancy for two statistical models is calculated as above, with the p_θ model substituted by one of the two models considered. Let $\Delta_Z(f, z_{\hat{\theta}})$ be the sample discrepancy function estimated for the model with density z_θ and let $\Delta_G(f, g_{\hat{\theta}})$ the sample discrepancy estimated for the model with density g_θ. Let us suppose that model g has a lower discrepancy, namely that $\Delta_Z(f, z_{\hat{\theta}}) = \Delta_G(f, g_{\hat{\theta}}) + \varepsilon$, where ε is a small positive number. Therefore, based on the comparison of the discrepancy functions we will choose the model with the density function g_θ.

This result may depend on the specific sample used to estimate the discrepancy function. We therefore need to carry out a statistical test to verify whether a discrepancy difference is significant; that is, whether the results obtained from a sample can be extended to all possible samples. If we find that the difference ε is not significant, then the two models would be considered equal and it would be natural to choose the simplest model. The deviance difference criterion defined by G^2 (Section 4.12.2) is equal to twice the difference between sample Kullback–Leibler discrepancies. For nested models, the G^2 difference is asymptotically equivalent to the chi-squared comparison (and test). When a Gaussian distribution is assumed, the Kullback–Leibler discrepancy coincides with the Euclidean discrepancy, therefore F statistics can also be used in this context.

To conclude, using a statistical test, it is possible to use the estimated discrepancy to make an accurate choice among the models. The disadvantage of this procedure is that it requires comparisons between model pairs, so when we have a large number of alternative models, we need to make heuristic choices regarding the comparison strategy (such as choosing among the forward, backward and stepwise criteria, whose results may diverge). Furthermore, we must assume a specific probability model and this may not always be a reasonable assumption.

5.2 Criteria based on scoring functions

In the previous section we have seen how the Kullback–Leibler sample discrepancy can be used to derive statistical tests to compare models. Often, however, we will not be able to derive a formal test. Examples can be found even among statistical models for data mining, for example models for data analysis with missing values or mixed graphical models. Furthermore, it may be important to have a complete ordering of models, rather than a partial one, based on pairwise comparisons. For this reason, it is important to develop scoring functions that assign a score to each model. The Kullback–Leibler discrepancy estimator is a scoring function that can often be approximated asymptotically for complex models.

A problem with the Kullback–Leibler score is that it depends on the complexity of a model, perhaps described by the number of parameters, hence its use may lead to complex models being chosen. Section 6.1 explained how a model selection strategy should reach a trade-off between model fit and model parsimony. We now look at this issue from a different perspective, based on a trade-off between bias and variance. In Section 4.9 we defined the mean squared error of an estimator. The mean squared error can be used to measure the Euclidean distance between the chosen model $p_{\hat{\theta}}$ and the underlying model f:

$$\mathrm{MSE}(p_{\hat{\theta}}) = E[(p_{\hat{\theta}} - f)^2].$$

Note that $p_{\hat{\theta}}$ is estimated on the basis of the data and is therefore subject to sampling variability. In particular, for $p_{\hat{\theta}}$ we can define an expected value $E(p_{\hat{\theta}})$, roughly corresponding to the arithmetic mean over a large number of repeated samples, and a variance $\mathrm{Var}(p_{\hat{\theta}})$, measuring its variability with respect to this expectation. From the properties of the mean squared error it follows that

$$\mathrm{MSE}(p_{\hat{\theta}}) = [\mathrm{bias}(p_{\hat{\theta}})]^2 + \mathrm{Var}(p_{\hat{\theta}}) = [E(p_{\hat{\theta}}) - f]^2 + E[(p_{\hat{\theta}} - E(p_{\hat{\theta}}))^2].$$

This indicates that the error associated with a model $p_{\hat{\theta}}$ can be decomposed into two parts: a systematic error (bias), which does not depend on the observed data, and reflects the error due to the parametric approximation; and a sampling error (variance), which reflects the error due to the estimation process. A model should therefore be selected to balance the two parts. A very simple model will have a small variance but a rather large bias (e.g. a constant model); a very complex model will have a small bias but a large variance. This is known as the bias–variance trade-off.

We now define score functions that penalise model complexity. The most important of these functions is the Akaike information criterion (AIC). Akaike (1974) formulated the idea that (i) the parametric model is estimated using the method of maximum likelihood and (ii) the parametric family specified contains the unknown distribution $f(x)$ as a particular case. He therefore defined a function that assigns a score to each model by taking a function of the Kullback–Leibler sample discrepancy. In formal terms, the AIC criterion is defined by the following equation:

$$\text{AIC} = -2 \log L(\hat{\theta}; x_1, \ldots, x_n) + 2q,$$

where $\log L(\hat{\theta}; x_1, \ldots, x_n)$ is the logarithm of the likelihood function calculated at the maximum likelihood parameter estimate and q is the number of parameters in the model. Notice that the AIC score essentially penalises the log-likelihood score with a term that increases linearly with model complexity.

The AIC criterion is based on the implicit assumption that q remains constant when the size of the sample increases. But this assumption is not always valid, so AIC does not lead to a consistent estimate for the dimension of the unknown model. An alternative and consistent scoring function is the Bayesian information criterion (BIC), formulated by Schwarz (1978) and defined by the following expression:

$$\text{BIC} = -2 \log L(\hat{\theta}; x_1, \ldots, x_n) + q \log(n).$$

It differs from the AIC criterion only in the second term, which now also depends on the sample size n. As n increases, BIC favours simpler models than AIC. As n grows large, the first term (linear in n) will dominate the second term (logarithmic in n). This corresponds to the fact that, for large n, the variance term in the MSE expression becomes negligible. Despite the superficial similarity between AIC and BIC, AIC is usually justified by resorting to classical asymptotic arguments, whereas BIC is usually justified by appealing to the Bayesian framework.

To conclude, the scoring function criteria we have examined are easy to calculate and lead to a total ordering of the models. Most statistical packages give the AIC and BIC scores for all the models considered. A further advantage of these criteria is that they can be used to compare non-nested models and, more generally, models that do not belong to the same class (e.g. a probabilistic neural network and a linear regression model). The disadvantage of these criteria is the lack of a threshold, as well as the difficulty of interpreting their measurement scale. In other words, it is not easy to determine whether or not the difference between two models is significant, and how it compares with another difference.

5.3 Bayesian criteria

From a practical perspective, the Bayesian criteria are an interesting compromise between the statistical criteria based on the deviance differences and the criteria based on scoring functions. They are based on coherent statistical modelling, and therefore their results can be easily interpreted. They provide a complete ordering of the models and can be used to compare non-nested models as well as models belonging to different classes. In the Bayesian derivation each model is given a score that corresponds to the posterior probability of the model itself. A model becomes a discrete random variable that takes values on the space of all candidate models. This probability can be calculated using Bayes' rule:

$$P(M|x_1, \ldots, x_n) = \frac{P(x_1, \ldots, x_n|M)P(M)}{P(x_1, \ldots, x_n)}.$$

The model that maximises the posterior probability will be chosen. Unlike the information criteria, the Bayesian criteria use probability and therefore define a distance that can easily be used to compare models. For further information about Bayesian theory and selection criteria for Bayesian models, see Bernardo and Smith (1994) and Cifarelli and Muliere (1989).

Bayesian scoring methods are not problem-free. Many Bayesian methods are hard to implement because of computational issues. For example, computing the likelihood of a model, $P(x_1, \ldots, x_n|M)$ can be a challenge, since it requires the parameters of the model to be integrated out. In other words, given a model M indexed by a vector θ of parameters, its likelihood is given by

$$P(x_1, \ldots, x_n|M) = \int P(x_1, \ldots, x_n|\theta, M)P(\theta|M)d\theta,$$

where $P(\theta|M)$ is the prior distribution of the parameters given that model M is under consideration. Although such calculations long prevented the widespread use of Bayesian methods, Markov chain Monte Carlo (MCMC) techniques emerged during the 1990s, providing a successful, albeit computationally intensive, way to approximate integration problems, even in highly complex settings. For a review of MCMC methods, see Gilks $et\ al.$ (1996). The most common software for implementing MCMC is BUGS, which can be found at www.mrc-bsu.cam.ac.uk/bugs.

On the other hand, Bayesian methods are quite attractive. Since the Bayesian model scores are probabilities, they can be used to draw model-averaged inferences from the various competing models, rather than making inferences conditional on a single model being chosen. Averaging across several models is a way to account for model uncertainty. Consider, for example, the problem of predicting the value of a certain variable Y. The Bayesian prediction, based on model averaging when there are K possible models, is given by

$$E(Y|x_1, \ldots, x_n) = \sum_{j=1}^{K} E(Y|M, x_1, \ldots, x_n)P(M|x_1, \ldots, x_n).$$

Notice how the prediction correctly reflects the uncertainty on the statistical model. Rather than choosing a single model, and drawing all inferences based on it, we can consider a plurality of models and average the inferences obtained from each model. The model average inference is a weighted average, where the weights are given by the model posterior probabilities. Application of Bayesian model averaging to complex models usually requires a careful implementation of computationally intensive techniques such as MCMC. This issue is considered in detail in Brooks *et al.* (2003) and, for graphical models, in Giudici and Green (1999). An important reference on Bayesian inference and computational approximations for highly structured stochastic systems is Green *et al.* (2003).

5.4   Computational criteria

The widespread use of computational methods has led to the development of computationally intensive model selection criteria. These criteria are usually based on using data sets that are different from the one being analysed (external validation) and are applicable to all the models considered, even when they belong to different classes (e.g. in comparing logistic regression, decision trees and neural networks, even when the latter two are non-probabilistic). A possible problem with these criteria is that they take a long time to design and implement, although general-purpose software such as R has made this task easier. We now consider the main computational criteria.

The cross-validation criterion

The idea of the cross-validation method is to divide the sample into two subsamples, a training sample having $n - m$ observations and a validation sample having m observations. The first sample is used to fit a model and the second is used to estimate the expected discrepancy or to assess a distance. We have already seen how to apply this criterion with reference to neural networks and decision trees. Using this criterion the choice between two or more models is made by evaluating an appropriate discrepancy function on the validation sample.

We can see that the logic of this criterion is different. The other criteria are all based on a function of internal discrepancy on a single data set, playing the roles of the training data set and the validation data set. With these criteria we compare directly predicted and observed values on an external validation sample. Notice that the cross-validation idea can be applied to the calculation of any distance function. For example, in the case of neural networks with quantitative output, we usually employ a Euclidean discrepancy,

$$\frac{1}{m} \sum_i \sum_j (t_{ij} - o_{ij})^2,$$

where t_{ij} is the fitted output and o_{ij} the observed output, for each observation i in the validation set and for each output neuron j.

One problem with the cross-validation criterion is in deciding how to select m, the number of the observations in the validation data set. For example, if we select $m = n/2$ then only $n/2$ observations are available to fit a model. We could reduce m but this would mean having few observations for the validation data set and therefore reducing the accuracy with which the choice between models is made. In practice, proportions of 75% and 25% are usually used for the training and validation data sets, respectively.

The cross-validation criterion can be improved in different ways. One limitation is that the validation data set is in fact also used to construct the model. Therefore the idea is to generalise what we have seen by dividing the sample in more than two data sets. The most frequently used method is to divide the data set into three blocks: training, validation and testing. The test data will not be used in the modelling phase. Model fit will be carried out on the training data, using the validation data to choose a model. Finally, the model chosen and estimated on the first two data sets will be adapted to the test set and the error found will provide a correct estimate of the prediction error. The disadvantage of this generalisation is that it reduces the amount of data available for training and validation.

A further improvement could be to use all the data available for training. The data is divided into k subsets of equal size; the model is fitted k times, leaving out one of the subsets each time, which could be used to calculate a prediction error rate. The final error is the arithmetic mean of the errors obtained. This method is known as k-fold cross-validation. Another common alternative is the leave-one-out method, in which one observation only is left out in each of the k samples, and this observation is used to calibrate the predictions. The disadvantage of these methods is the need to retrain the model several times, which can be computationally intensive.

The bootstrap criterion

The bootstrap method was introduced by Efron (1979) and is based on the idea of reproducing the 'real' distribution of the population with a resampling of the observed sample. Application of the method is based on the assumption that the observed sample is in fact a population, a population for which we can calculate the underlying model $f(x)$ – it is the sample density. To compare alternative models, a sample can be drawn (or resampled) from the fictitious population (the available sample) and then we can use our earlier results on model comparison. For instance, we can calculate the Kullback–Leibler discrepancy directly, without resorting to estimators. The problem is that the results depend on the resampling variability. To get around this, we resample many times, and we assess the discrepancy by taking the mean of the obtained results. It can be shown that the expected discrepancy calculated in this way is a consistent estimator of the expected discrepancy of the real population.

Application of the bootstrap method requires the assumption of a probability model, either parametric or non-parametric, and tends to be computationally intensive.

Bagging and boosting

Bootstrap methods can be used not only to assess model's discrepancy, and therefore its accuracy, but also to improve the accuracy. Bagging and boosting methods are recent developments that can be used for combining the results of more than one data mining analysis. In this respect they are similar to Bayesian model-averaging methods, as they also lead to model-averaged estimators, which often improve on estimators derived from only one model.

Bagging (bootstrap aggregation) methods can be described as follows. At every iteration, we draw a sample with replacement from the available training data set. Typically, the sample size corresponds to the size of the training data itself. This does not mean that the sample drawn will be the same as the training sample, because observations are drawn with replacement – not all the observations in the original sample are drawn. Consider B loops of the procedure; the value of B depends on the computational resources available and time. A data mining method can be applied to each bootstrapped sample, leading to a set of estimates for each model; these can then be combined to obtain a bagged estimate. For instance, the optimal classification tree can be searched for each sample, and each observation allocated to the class with the highest probability. The procedure is repeated, for each sample $i = 1, \ldots, B$, leading to B classifications. The bagged classification for an observation corresponds to the majority vote, namely, to the class in which it is most classified by the B fitted trees. Similarly, a regression tree can be fitted for each of the B samples, producing a fitted value $\hat{y}_i$, in each of them, for each observation. The bagged estimate would be the mean of these fitted values,

$$\frac{1}{B} \sum_{i=1}^{B} \hat{y}_i.$$

With reference to the bias–variance trade-off, as a bagged estimate is a sample mean, it will not alter the bias of a model; however, it may reduce the variance. This can occur for highly unstable models, such as decision trees, complex neural networks and nearest-neighbour models. On the other hand, if the applied model is simple, the variance may not decrease, because the bootstrap variability dominates.

So far we have assumed that the same model is applied to the bootstrap samples; this need not be the case. Different models can be combined, provided the estimates are compatible and expressed on the same scale. While bagging relies on bootstrap samples, boosting does not. Although now there are many variants, the early versions of boosting fitted models on several weighted versions of the data set, where the observations with the poorest fit receive the greatest weight. For instance, in a classification problem, the well-classified observations will get lower weights as the iteration proceeds, allowing the model to concentrate on the estimating the most difficult cases. More details can be found in Han and Kamber (2001) and Hastie *et al.* (2001).

5.5 Criteria based on loss functions

One aspect of data mining is the need to communicate the final results in accordance with the aims of the analysis. With business data we need to evaluate models not only by comparing them among themselves but also by comparing the business advantages which can be gained by using one model rather than another. Since the main problem dealt with by data analysis is to reduce uncertaintiesin the risk factors or loss factors, we often talk about developing criteria that minimise the loss connected with a problem. In other words, the best model is the one that leads to the least loss. The best way to introduce these rather specific criteria is to give some examples. Since these criteria are mostly used in predictive classification problems, we will mainly refer to this context here.

The confusion matrix is used as an indication of the properties of a classification (discriminant) rule (see the example in Table 5.1). It contains the number of elements that have been correctly or incorrectly classified for each class. The main diagonal shows the number of observations that have been correctly classified for each class; the off-diagonal elements indicate the number of observations that have been incorrectly classified. If it is assumed, explicitly or implicitly, that each incorrect classification has the same cost, the proportion of incorrect classifications over the total number of classifications is called the error rate, or misclassification error; this is the quantity we must minimise. The assumption of equal costs can be relaxed by weighting errors with their relative costs.

We now consider the lift chart, and the ROC curve, two graphs that can be used to assess model costs. Both are presented with reference to a binary response variable, the area where evaluation methods have developed most quickly. For a comprehensive review, see Hand (1997).

Lift chart
The lift chart puts the observations in the validation data set into increasing or decreasing order on the basis of their score, which is the probability of the response event (success), as estimated on the basis of the training set. It groups these scores into deciles, then calculates and graphs the observed probability of success for each of the decile classes in the validation data set. A model is valid if the observed success probabilities follow the same order (increasing or

Table 5.1 Example of a confusion matrix.

	Observed classes		
Predicted classes	Class A	Class B	Class C
Class A	45	2	3
Class B	10	38	2
Class C	4	6	40

Table 5.2   Theoretical confusion matrix.

Predicted Observed	Event (1)	Non-event (0)	Total
Event (1)	a	b	$a+b$
Non-event (0)	c	d	$c+d$
TOTAL	$a+c$	$b+d$	$a+b+c+d$

decreasing) as the estimated probabilities. To improve interpretation, a model's lift chart is usually compared with a baseline curve, for which the probability estimates are drawn in the absence of a model, that is, by taking the mean of the observed success probabilities.

ROC curve

The receiver operating characteristic (ROC) curve is a graph that also measures predictive accuracy of a model. It is based on the confusion matrix in Table 5.2. In the table, the term 'event' stands for the value $Y = 1$ (success) of the binary response. The confusion matrix classifies the observations of a validation data set into four possible categories:

- observations correctly predicted as events (with absolute frequency equal to a);
- observations incorrectly predicted as events (with frequency equal to c);
- observations incorrectly predicted as non-events (with frequency equal to b);
- observations correctly predicted as non-events (with frequency equal to d).

Given an observed table, and a cut-off point, the ROC curve is calculated on the basis of the resulting joint frequencies of predicted and observed events (successes) and non-events (failures). More precisely, it is based on the following conditional probabilities:

- *sensitivity*, $a/(a + b)$, the proportion of events predicted as such;
- *specificity*, $d/(c + d)$, the proportion of non-events predicted as such;
- *false positives*, $c/(c + d) = 1 -$ specificity, the proportion of non-events predicted as events (type II error);
- *false negatives*, $b/(a + b) = 1 -$ sensitivity, the proportion of events predicted as non-events (type I error).

The ROC curve is obtained by graphing, for any fixed cut-off value, the false positives on the horizontal axis and the sensitivity on the vertical axis (see Figure 5.1

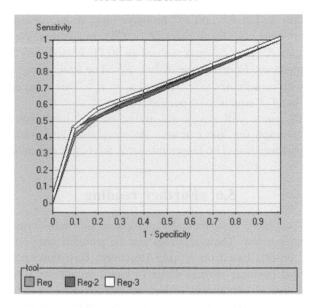

Figure 5.1 Example of an ROC curve.

for an example). Each point in the curve corresponds to a particular cut-off. The ROC curve can also be used to select a cut-off point, trading off sensitivity and specificity. In terms of model comparison, the ideal curve coincides with the vertical axis, so the best curve is the leftmost curve.

The ROC curve is the basis for an important summary statistic called the Gini index of performance. Recall the concentration curve in Figure 3.2. For any given value of F_i, the cumulative frequency, there is a corresponding value of Q_i, the cumulative intensity. F_i and Q_i take values in [0,1] and $Q_i \leq F_i$. The concentration curve joins a number of points in the Cartesian plane determined by taking $x_i = F_i$ and $y_i = Q_i$, for $i = 1, \ldots, n$. The area between the curve and the 45° line gives a summary measure for the degree of concentration. The ROC curve can be treated in a similar way. In place of F_i and Q_i we need to consider two cumulative distributions constructed as follows.

The data contains both events ($Y_i = 1$) and non-events ($Y_i = 0$). It can therefore be divided into two samples, one containing all events (labelled E) and one containing all non-events (labelled N). As we have seen, any statistical model for predictive classification takes each observation and attaches to it a score that is the fitted probability of success π_i. In each of the two samples, E and N, the observations can be ordered (in increasing order) according to this score. Now, for any fixed value of i (a percentile corresponding to the cut-off threshold), a classification model would consider all observations below it as non-events and all observations above it as events.

Correspondingly, the predicted proportion of events can be estimated for both E and N. For a reasonable model, in population E this proportion has to be higher than in population N. Let F_i^E and F_i^N be these proportions corresponding to the cut-off i, and calculate coordinate pairs (F_i^E, F_i^N), as i varies. We have that, for $i = 1, \ldots, n$, both F_i^E and F_i^N take values in [0,1]; indeed they both represent cumulative frequencies. Furthermore, $F_i^N \leq F_i^E$. The ROC curve is obtained by joining points with coordinates $y_i = F_i^E$ and $x_i = F_i^N$. This is because F_i^E equals the sensitivity and F_i^N equals 1 – specificity.

Notice that the curve will always lie above the 45° line. However, the area between the curve and the line can also be calculated, and coincides with the Gini index of performance. The larger the area, the better the model.

5.6   Further reading

In this chapter we have systematically compared the main criteria for model selection and comparison. These methods can be grouped into: criteria based on statistical tests, criteria based on scoring functions, Bayesian criteria, computational criteria, and business criteria. Criteria based on statistical tests start from the theory of statistical hypothesis testing, so there is a lot of detailed literature related to this topic; see, for example, Mood *et al.* (1991). The main limitation of these methods is that the choice among the different models is made by pairwise comparisons, thus leading to a partial ordering.

Criteria based on scoring functions offer an interesting alternative, since they can be applied in many settings and provide a complete ordering of the models. In addition, they can be easily computed. However, they do not provide threshold levels for assessing whether the difference in scores between two models is significant. Therefore they tend to be used in the exploration or preliminary phase of the analysis. For more details on these criteria and how they compare with the hypothesis testing criteria, see Zucchini (2000) or Hand *et al.* (2001). Bayesian criteria are a possible compromise between the previous two. However, Bayesian criteria are not widely used, since they are not implemented in the most popular statistical software. For data mining case-studies that use Bayesian criteria, see Giudici (2001) and Giudici and Castelo (2001).

Computational criteria have the advantage that they can be applied to statistical methods that are not necessarily 'model based'. From this point of view they are the main principle of 'universal' comparison among the different types of models. On the other hand, since most of them are non-probabilistic, they may be too dependent on the sample observed. A way to overcome this problem is to consider model combination methods, such as bagging and boosting. For a thorough description of these recent methodologies, see Hastie *et al.* (2001).

Criteria based on loss functions are relatively recent, even though the underlying ideas have been used in Bayesian decision theory for quite some time; see Bernardo and Smith (1994). They are of great interest and have great application potential, even though presently they are used only in the context of classification. For a more detailed examination of these criteria, see Hand (1997), Hand *et al.* (2001), or the manuals for the R statistical software.

PART II

Business case studies

CHAPTER 6

Describing website visitors

6.1 Objectives of the analysis

In this chapter we present an analysis of web access data, the main objective of which is to classify the visitors into homogeneous groups on the basis of their behaviour. This will lead us to identify typical visiting profiles. In other words, we are trying to match each visitor to a specific cluster, depending on their surfing habits on that particular site. This will give us a behavioural segmentation of the users that we can use in future marketing decisions. We can also monitor the evolution of the kind of 'customer' who comes to the site by looking at how the distribution of users in the different behavioural segments evolves over time. Then we can assess the impact of any changes to the site (such as reorganisation or restyling of web pages, advertisements or promotions) on the number of the visits of the different types of users. The data set used in this chapter has also been analysed by Cadez *et al.* (2000) and, from a Bayesian perspective, by Giudici and Castelo (2001).

6.2 Description of the data

The data set contains data about the pages of the site www.microsoft.com visited by 32 711 anonymous visitors. For each visitor we have indicated the pages of the site that have been visited during the first week of February 1998. Visitors are assigned an identification number (from 10 001 to 42 711) and no personal information is given. The total number of visited pages is 296. The pages are identified by a number that corresponds to a title and a corresponding address. For example, number 1057 refers to the page 'MS PowerPoint News' of the PowerPoint group of pages. The numeric codes associated with the pages are integers that go from 1000 up to 1295. To give a better idea of the data set, here are its first few lines:

```
C, ''10908'', 10908
V, 1108
```

Applied Data Mining for Business and Industry, 2e   P. Giudici, S. Figini
© 2009 John Wiley & Sons, Ltd

```
V, 1017
C, ''10909'', 10909
V, 1113
V, 1009
V, 1034
C, ''10910'', 10910
V, 1026
V, 1017
```

Each visitor is represented in the data set by a line (beginning with the letter C) containing their identification number. The code is converted into a number. The visitor's line is then followed by one or more lines that show the pages visited. For the objectives of the analysis it is convenient to work with a derived data matrix, organised by visitors. This matrix will describe, for each visitor, how many times each page has been viewed; therefore every page will be represented by one categorical variable.

Since the total number of distinct pages in the database is 294, it is likely that some combinations of these pages are never or rarely visited. To perform a valid cluster analysis of visitors into groups, it is therefore useful to clean and summarize the original file so as to obtain a less complex data matrix. To do so, we group the web pages into 13 homogeneous categories, reflecting their meaning in the Microsoft website. The number of variables in the data set is thus reduced from 294 to 13. Each variable corresponds to one of the 13 groups:

- **Initial.** this includes all the general access pages and all the pages dedicated to research.
- **Support.** this includes all the pages related to requests for help and support.
- **Entertainment.** this includes all the pages that refer to entertainment, games and cultural software.
- **Office.** this has all the pages that refer to the Office software.
- **Windows.** this groups together all the pages related to the Windows operating system.
- **Othersoft.** this refers to all the pages relating to software other than Office.
- **Download.** this includes all the pages regarding software downloading or updating.
- **Otherint.** this has all the pages dedicated to services through internet, for information technology professionals, which are different from the download pages.
- **Development.** this has all the pages dedicated to professional developers (e.g. Java).
- **Hardware.** this includes the pages relating to Microsoft hardware.
- **Business.** this has pages dedicated to businesses.
- **Info.** this includes all the pages which give information about new products and services.
- **Area.** this has all the pages which refer to local access, depending on the specific language.

Table 6.1 Data structure.

Client_Code	Initial	Help	Entertainment
10001	1	1	1
10002	1	1	0
10003	2	1	0
10004	0	0	0
10005	0	0	0
10006	2	0	0
10007	0	0	0
10008	1	0	0
10009	0	0	0
10010	1	1	0
10011	2	0	0
10012	0	0	0
10013	0	0	0

Using this grouping, we can derive a visitor data matrix with 32 711 rows and 13 columns. Table 6.1 shows part of it. Every group of pages is represented by a discrete variable that counts the number of times each person has visited that specific group of pages. The data set does not include information on the order in which the pages are visited.

We carry out the analysis in three stages: first, an exploratory phase; then we determine the behavioural classes of users using descriptive data mining techniques (cluster analysis and Kohonen maps); and finally, we compare the performance of the two descriptive models.

6.3 Exploratory analysis

The exploratory analysis of the data reveals a high level of dispersion. Table 6.2 shows the absolute frequency distribution of the number of times each group of pages is visited. Notice that some groups of pages (such as Initial, Office and Download) have a high frequency of visits; others (such as Information, Business, Hardware and Entertainment) have a much lower frequency. We point out that, on average, a user visits 4 distinct pages. However, the modal number of visits is only 2, indicating that the variable 'number of visited pages' is positively skewed.

6.4 Model building

The first part of the analysis aims to identify the different behavioural segments within the sample of users. We use two different descriptive data mining techniques: cluster analysis and the unsupervised networks known as Kohonen maps. Both techniques allow us to partition the data to identify homogeneous groups or types possessing internal cohesion that differentiates them from the other groups.

Table 6.2 Frequency distribution.

Page	Frequency
Initial	23492
Help	9287
Entertainment	2967
Office	15574
Windows	7328
Othersft	3046
Download	11320
Otherint	6237
Devolpment	8228
Hardware	2967
Business	2726
Information	2307
Area	3141

We use two techniques so we can compare their efficiency, but also to check that they produce consistent results.

6.4.1 Cluster analysis

Chapter 4 explained the main techniques of hierarchical cluster analysis as well as the non-hierarchical K-means method. A variation of the K-means method is used here. The basic idea is to introduce seeds, or centroids, to which statistical units may be attracted, forming a cluster. It is important to specify the maximum number of clusters, say G, in advance. As discussed in Section 4.2, hierarchical and non-hierarchical methods of cluster analysis do have some disadvantages. Hierarchical cluster analysis does not need to know the number of clusters in advance, but it may require too much computing power. For moderately large quantities of data, as in this case study, the calculations may take a long time. Non-hierarchical methods are fast, but they require us to choose the number of clusters in advance.

To avoid these disadvantages and to try to exploit the potential of both methods we follow a combined approach. First we run a non-hierarchical clustering procedure on the entire data set, having chosen a large value of G. We take the first G available observations as seeds. Then we run an iterative procedure; at each step we form temporary clusters, allocating each observation to the cluster with the seed nearest to it. Each time an observation is allocated to a cluster, the seed is substituted with the mean of the cluster – the centroid – itself. We repeat the iterative process until convergence; that is, until no substantial changes in the cluster seeds are evident. At the end of the procedure, we have G clusters, with corresponding centroids.

This is the input to the next step, a hierarchical clustering procedure on a sample from the available data, the aim of which is to find the optimal number of clusters. The procedure is of course an agglomerative one, since the number of clusters cannot be greater than G.

Having ascertained the optimal number of clusters, we carry out a non-hierarchical clustering procedure to allocate the observations to the clusters, whose initial seeds are the centroids obtained in the previous step. The procedure is similar to the first non-hierarchical stage, and involves repeating the following two steps until convergence:

1. Scan the data and assign each observation to the seed that is nearest (in terms of Euclidean distance).
2. Replace each seed with the mean of the observations assigned to its cluster.

Here we choose $G = 40$. We carry out the hierarchical stage of the procedure on a sample of 2000 observations from the available data. Our distance function is the Euclidean distance, and we use Ward's method to recompute the distances as the clusters are formed. To obtain valid cluster means for use as seeds in the third stage, we impose a minimum of 100 observations in each cluster.

By applying Ward's method, we obtain that the optimal number of clusters is 6. Applying the centroid method gives the same result. Running a final non-hierarchical procedure on the entire available data set, with six clusters, gave the results presented in Table 6.3. This shows the number of observations in each cluster. We have $R^2 = 0.40$ for the final configuration, which can be treated as a summary evaluation measure of the model.

To better interpret the cluster configurations, Table 6.4 gives the means of each cluster for the most important variables. Note that clusters 1 and 6 have similar centroids, expressed by a similar mean number of visits to each page (especially Office, Entertainment and Windows). On the other hand, cluster 2 appears to have rather different behaviour, concentrated mainly on three pages (Help, Office and Windows).

6.4.2 Kohonen networks

Kohonen networks require us to specify the number of rows and the number of columns in the grid space characterising the map. Large maps are usually the

Table 6.3 Cluster sizes for the final K-means cluster configuration.

Cluster	Frequency
1	10725
2	60
3	19277
4	164
5	2325
6	160

Table 6.4   Cluster means for the final K-means cluster configuration.

Web page	Cluster 1	Cluster 2	Cluster 3	Cluster 4	Cluster 5	Cluster 6
Help	0.25	2.41	0.26	0.70	0.44	0.61
Download	1.01	0.91	0	0.75	0.07	0.41
Office	0.70	1.63	0.26	1.14	1.11	0.70
Entertainment	0.10	0.75	0.08	0.18	0.08	0.13
Windows	0.30	1.7	0.06	1.64	1.02	0.26

best choice, as long as each cluster has a significant number of observations. The learning time increases significantly with the size of the map. The number of rows and the number of columns are usually established by conducting several trials until a satisfactory result is obtained. We will use the results of the cluster analysis to help us. Having identified 6 as the optimal number of clusters, we will consider a 3×2 map. The Kohonen mapping algorithm implemented in R essentially replaces the third step of the clustering algorithm with a procedure that repeats the following two steps until convergence:

1. Scan the data and assign each observation to the seed that is nearest (in terms of Euclidean distance).
2. Replace each seed with a weighted mean of the cluster means that lie in the grid neighbourhood of the seed's cluster.

The weights correspond to the frequencies of each cluster. In this way the cluster configuration is such that any two clusters that are close to each other in the map grid will have centroids close to each other. The initial choice of the seeds can be made in different ways; we choose them at random. Alternatively, we could have used the centroids obtained from the second stage of the K-means clustering procedure.

Table 6.5 reports, for each of the six chosen map clusters, the total number of observations in it (frequency). The groups obtained are now more homogeneous in terms of number of observations included. Table 6.6, which reports the cluster means, should be compared with Table 6.4 for the K-means procedure. R^2 is now

Table 6.5   Cluster sizes for the final Kohonen map configuration.

Cluster	Frequency
1	9572
2	5784
3	8301
4	1863
5	4995
6	2196

Table 6.6 Cluster means for the final Kohonen map configuration.

Web page	Cluster 1	Cluster 2	Cluster 3	Cluster 4	Cluster 5	Cluster 6
Help	0.40	0.42	0.40	0.49	0.90	0.52
Download	0.64	0.43	0.39	0.44	0.49	0.48
Office	0.67	0.42	0.38	0.43	0.42	0.61
Entertainment	0.46	0.47	0.54	0.49	0.50	0.51
Windows	0.47	0.45	0.51	0.49	0.56	0.51

0.58, which is 0.18 higher than we obtained for the K-means procedure. From Table 6.6 we conclude that the findings in Table 6.4 are substantially confirmed.

6.5   Model comparison

We have presented two ways to perform descriptive data mining on web data. Broadly speaking, a visitor profile is better if the cluster profiles are more distinct and if their separation reflects a truly distinct behaviour. The Kohonen map does seem to perform better in this respect, by exploiting the dependence between adjacent clusters.

A second consideration is that the statistical evaluation of the results should be based mainly on R^2, or measures derived from it, this being a descriptive analysis. We have already seen that the overall R^2 is larger with the Kohonen networks. It is interesting to examine for each variable (page) the ratio of the between sum of squares and the total sums of squares that leads to R^2. This can give a measure of the goodness of fit of the cluster representation, specific for each variable. By examining all such R^2 we can get an overall picture of which aspects of the observations are more used in the clustering process.

Table 6.7 presents the variable-specific R^2 values and the overall R^2 for the K-means and Kohonen procedures. For both procedures the group pages that have a high R^2, and are therefore most influential in determining the final results, are Help and Office. There are also pages that are influential only for the Kohonen maps: Windows and Download. The choice between the two procedures therefore depends on which pages are chosen as discriminant for the behaviour. In the

Table 6.7 Comparison of the variable-specific R^2 values.

Web page	R^2 (K-means)	R^2 (Kohonen)
Help	0.68	0.66
Entertainment	0.01	0.02
Office	0.44	0.70
Windows	0.18	0.56
Download	0.04	0.43

absence of other considerations, the choice should consider the procedure which leads to the highest overall R^2, here the Kohonen map.

Further considerations may arise when the results are used to make predictions. For instance, suppose that, once the grouping has been accomplished, we receive some new observations to be classified into clusters. One reasonable way to proceed is to assign them to one of the clusters previously determined, according to a discriminant rule. In that case, clustering methods can be compared in terms of predictive performance, perhaps by using cross-validation techniques.

6.6   Summary report

1. **Context.** This case is concerned with customer profiling on the basis of web behaviour. The context is very broad, as the analysis refers to any type of problem involved with classifying people, companies or any other statistical units into homogeneous groups.
2. **Objectives.** The aim of the analysis is to classify customers into an unknown number of homogeneous classes, on the basis of their statistical characteristics. The classification is unsupervised: there is no target variable and all the available information should be used to form homogeneous clusters.
3. **Organisation of the data.** The data was extracted from a log file that registers access to a website. The data matrix records for each visitor the number of times they have viewed a collection of pages. For computational tractability, the pages were grouped into 13 web areas, homogeneous in terms of their content. Therefore, the data matrix considered in the analysis contains 32 711 rows (visitors) and 13 columns (one counting variable for each area).
4. **Exploratory data analysis.** This phase of the analysis revealed a high level of dispersion with respect to the pages visited. Each visitor looks, on average, at 4 pages, and this confirms the validity of grouping the 104 visited pages into 13 areas.
5. **Model specification.** The objective of the analysis was to group the observations into homogeneous classes. Given the size of the data set, we considered non-hierarchical cluster analysis models based on the K-means algorithm and Kohonen networks. To compare the two approaches fairly, we considered a 3×2 Kohonen map, which corresponds to the same number of clusters (6) obtained with the K-means algorithm.
6. **Model comparison.** Models were first compared by splitting the total variability into within-group variability and between-group variability, leading to the calculation of R^2, both overall and for specific area variables. The result of the comparison favours Kohonen networks, which also have the advantage that the groups obtained tend to be more distinct than the groups from K-means clustering. We then compared the models in terms of their predictive ability. We did this by using the clustering variable as a 'target' variable, fitting a classification tree and following a cross-validation approach. The results again favour Kohonen networks.

7. **Model interpretation.** The interpretation of the results should be based on the cluster profiles obtained. For the Kohonen map, which performed best, we interpreted each cluster profile by looking at the comparison between the cluster-specific mean of each of the 13 variables and the overall mean. Expert knowledge is needed to elucidate the business meaning of each profile.

CHAPTER 7

Market basket analysis

7.1 Objectives of the analysis

This case study looks at market consumer behaviour using a marketing methodology known as market basket analysis. Market basket analysis has the objective of indentifying products, or groups of products, that tend to occur together (are associated) in buying transactions (baskets). The knowledge obtained from a market basket analysis can be very valuable; for instance, it can be employed by a supermarket to reorganise its layout, taking products frequently sold together and locating them in close proximity. But it can also be used to improve the efficiency of a promotional campaign: products that are associated should not be put on promotion at the same time. By promoting just one of the associated products, it should be possible to increase the sales of that product and get accompanying sales increases for the associated products.

The databases usually considered in a market basket analysis consist of all the transactions made in a certain sale period (e.g. one year) and in certain sale locations (e.g. a chain of supermarkets). Consumers can appear more than once in the database. In fact, consumers will appear in the database whenever they carry out a transaction at a sales location. The objective of the analysis is to find the most frequent combinations of products bought by the customers. The association rules in Section 4.8 represent the most natural methodology here; indeed they were actually developed for this purpose. Analysing the combinations of products bought by the customers, and the number of times these combinations are repeated, leads to a rule of the type 'if condition, then result' with a corresponding interestingness measurement. Each rule of this type describes a particular local pattern. The set of association rules can be easily interpreted and communicated. Possible disadvantages are locality and lack of probability modelling.

This case study takes a real market basket analysis and compares association rules with log-linear models (Section 4.13), which represent a powerful method of descriptive data mining. It also shows how an exploratory analysis, based on examining the pairwise odds ratios, can help in constructing a comprehensive log-linear model. Odds ratios can be directly compared with association rules.

Applied Data Mining for Business and Industry, 2e P. Giudici, S. Figini
© 2009 John Wiley & Sons, Ltd

Similar analyses can be found in Giudici and Passerone (2002) and Castelo and Giudici (2003); Castelo and Giudici take a Bayesian viewpoint.

7.2 Description of the data

The statistical analysis in this chapter was carried out on a data set kindly provided by AC Nielsen, concerning transactions at a large supermarket in southern Italy. The data set is part of a larger database for 37 shop locations of a chain of supermarkets in Italy. In each shop the recorded transactions are all the transactions made by someone holding one of the chain's loyalty cards. Each card carries a code that identifies features about the owner, including important personal characteristics such as sex, date of birth, partner's date of birth, number of children, profession and education. The card allows the analyst to follow the buying behaviour of its owner: how many times they go to the supermarket in a given period, what they buy, whether they follow the promotions, etc. Our aim here is to consider only data on products purchased, in order to investigate the associations between these products. Therefore we shall not consider the influence of demographic variables or the effect of promotions.

The available data set is organised in a collection of 37 transactional databases, one for each shop location. For each shop, a statistical unit (a row in the database) corresponds to one loyalty card code and one product bought. For each card code there may be more than one product and, in the file, the same card code may appear more than once, each time corresponding to one visit to a particular shop. The period considered consists of 75 days between 2 January and 21 April 2001. To suit the aims of the analysis and the complexity of the overall data set, we will choose one representative shop, in southern Italy, with an area of about 12 000 m^2. This shop has a mean number of visits, in the period considered, of 7.85, which is roughly equivalent to the overall mean for the 37 shops. But the total number of loyalty cards issued for the shop is 7301, the largest out of all the shops; this is one of the main reasons for choosing it. Finally, the average expenditure per transaction is about €28.27, slightly lower than the overall mean.

The total number of products available in the shop is about 5000, ignoring the brand, format and specific type (e.g. weight, colour, size). Products are usually grouped into categories. The total number of available categories in the supermarket considered is about 493. For clarity we will limit our analysis to 20 categories (items), corresponding to those most sold. They are listed in Table 7.1, along with their frequency of occurrence, namely, the number of transactions that contain the item at least once; Figure 7.1 provides a graphical display of Table 7.1. Notice that all product categories considered – shortened to products from now on – concern food products. These categories are used to produce a transaction database and Table 7.2 presents an extract. This extract will be called the transactions data set.

From Table 7.2 notice that the transaction database presents, for each card transaction, a list of the products that have been put in the basket. For example, card owner 0460202004099 has bought tinned meat, tuna and mozzarella. The

Table 7.1 Frequency of occurrence.

Product	Frequency
beer	2082
biscuits	4863
brioches	3954
coffee	3044
coke	2098
crackers	557
frozen fish	636
frozen vegetables	2959
ice cream	222
juices	2126
milk	10999
mozzarella	317
oil	661
pasta	14707
rice	1481
tinned meat	122
tomato sauce	2484
tuna	2034
water	6420
yoghurt	3769

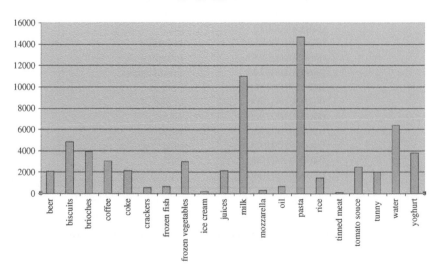

Figure 7.1 Graphical frequency distribution.

transaction database may be expressed as a data matrix, with each row representing one transaction by one owner of a card code; the columns are binary variables that represent whether or not each specific product has been bought (at least once) in that transaction. We will call this the card owners database;

Table 7.2 Data structure.

COD_CARD	Product
0460202004099	tinned meat
0460202004099	tuna
0460202004099	mozzarella
0460202007021	milk
0460202033648	milk
0460202033648	coke
0460202033648	pasta
0460202033648	crackers
0460202033648	milk
0460202035871	water
0460202035871	water
0460202039190	tuna
0460501000020	crackers
0460501000020	milk
0460501000020	pasta
0460501000020	biscuits
0460501000020	pasta
0460501000020	biscuits
0460501000204	juices
0460501000204	milk
0460501000204	biscuits
0460501000303	pasta
0460501000723	oil
0460501000853	biscuits
0460501000853	biscuits
0460501001744	tinned meat
0460501001966	milk
0460501001966	coffee
0460501001966	pasta
0460501001966	pasta
0460501001980	tuna
0460501001980	milk
0460501001980	tomato sauce

and extract is shown in Table 7.3. Note that the total number of card owner transactions is 46 727.

7.3   Exploratory data analysis

To understand the associations between the 20 products considered, we have considered 190 two-way contingency tables, one for each pair of products. Table 7.4 shows one of these tables. It can be used to study the association between the products ice cream and Coke. In each cell of the contingency table we have

Table 7.3 The card owners database.

Cod_Cart	Tin_Meat	Mozzar	Tuna	Milk	Coke	Pasta
460202004099	0	1	0	1	0	0
460202004098	0	0	0	1	0	1
460202004097	0	0	0	0	0	0
460202004096	1	0	0	0	0	1
460202004095	0	1	0	0	1	0
460202004094	0	1	0	1	1	1
460202004093	0	1	0	1	1	0
460202004092	0	1	0	1	1	0
460202004091	0	1	0	1	1	0

Table 7.4 Example of a two-way contingency table, and calculation of the odds ratios.

ICECREAM   COKE

Frequency Percent Row Pct Col Pct	0	1	Total
0	41179 88.13 89.60 98.57	4779 10.23 10.40 96.56	45958 98.35
1	599 1.28 77.89 1.43	170 0.36 22.11 3.44	769 1.65
Total	41778 89.41	4949 10.59	46727 100.00

	Value	95% Confidence Limits	
Odds Ratio	2.4455	2.0571	2.9071

the absolute frequency, the relative frequency (as a percentage), and the conditional frequency by row and by column. Below the table we report an association measure, the odds ratio between the two variables, along with the corresponding confidence interval. According to Section 3.4, an association is deemed significant if the value 1 is outside the confidence interval. Here we can say there is a strong positive association between the two products. Recall that the total sample size is quite large (46 727 transactions), therefore even a small odds ratio can be significant. We have calculated all 190 possible odds ratios between products; the largest values are shown in Table 7.5. Notice that the largest associations are detected between tinned meat and tuna, tinned meat and mozzarella, and

Table 7.5 The largest odds ratios between pairs of products, and the corresponding confidence interval.

Product 1	Product 2	Odds ratio	Confidence interval
BRIOCHES	CRACKERS	2.28	2.02, 2.57
BRIOCHES	ICE CREAM	2.02	1.71, 3.37
BRIOCHES	JUICES	2.80	2.60, 3.02
COKE	BEER	2.81	2.61, 3.02
COKE	ICE CREAM	2.44	2.05, 2.90
CRACKERS	ICE CREAM	2.28	1.70, 3.05
CRACKERS	JUICES	2.04	1.76, 2.38
FROZEN VEGETABLES	FROZEN FISH	3.36	2.95, 3.82
FROZEN FISH	MOZZARELLA	2.07	1.47, 2.93
JUICES	ICE CREAM	2.53	2.10, 3.05
OIL	TOMATO SAUCE	2.07	1.83, 2.34
RICE	PASTA	2.11	1.96, 2.27
TINNED MEAT	RICE	2.14	1.47, 3.11
TINNED MEAT	TUNA	5.06	3.91, 6.56
TINNED MEAT	MOZZARELLA	4.88	2.96, 8.03
TOMATO SAUCE	TUNA	5.06	3.91, 6.56

frozen fish and frozen vegetables. In all these cases the two paired products are fast food products. Next comes an association between two drinks: Coke and beer. In general, all the associations in Table 7.5 appear fairly reasonable from a subject-matter viewpoint. In calculating the odds ratios, each pair of variables is considered independently of the remaining 18. It is possible to relate them to each other by drawing a graph whose the nodes are the products. An edge is drawn between a pair of nodes if the corresponding odds ratio is significantly different from 1; in other words, if the confidence interval for the odds ratio does not contain the value 1.

In Table 7.6 we report for each pair of products whether the association between them is positive, negative or absent. We can then proceed by grouping together linked products. Five products appear to be isolated from the others, not (strongly) positively associated with anything: milk, biscuits, water, coffee and yoghurt. All other products are related, either directly or indirectly. It is possible to find at least three groups of connected groups. These three groups indentify typical buying behaviours. There is one group with five nodes: tuna, tinned meat, mozzarella, frozen fish and frozen vegetables. These nodes, highly related with each other, correspond to fast food products, quick and easy to prepare. A second group contains four nodes: rice, pasta, tomato sauce and oil. This group can be identified with food bought for ordinary meals (ordinary by Mediterranean standards). A third group contains six other products: beer, Coke, juice, ice cream, brioches and crackers. All relate to break items, food and drink typically consumed outside of regular meals. This group seems less logically homogeneous than the other two. We shall return to this in the next section. We have not detected any significant

Table 7.6 Association between products.

Product 1	Product 2	Association
BRIOCHES	CRACKERS	+
BRIOCHES	ICE CREAM	+
BRIOCHES	JUICES	+
COKE	BEER	+
COKE	ICE CREAM	+
CRACKERS	ICE CREAM	+
CRACKERS	JUICES	+
FROZEN VEGETABLES	FROZEN FISH	+
FROZEN FISH	MOZZARELLA	+
JUICES	ICE CREAM	+
OIL	TOMATO SAUCE	+
RICE	PASTA	+
TINNED MEAT	RICE	+
TINNED MEAT	TUNA	+
TINNED MEAT	MOZZARELLA	+
TOMATO SAUCE	TUNA	+

negative association in the data at hand. This has important implications; for instance, a promotion on pasta will presumably increase the sales of this product but is very unlikely to decrease the sales of other products, such as rice and water. Negative associations are rarely considered in market basket analysis.

7.4 Model building

7.4.1 Log-linear models

Log-linear models are very useful for descriptive data mining; they investigate the associations between the variables considered. Fitting a log-linear model to all 20 of our binary variables may require too many parameters to be estimated. Furthermore, the corresponding conditional independence graph may be difficult to interpret. Therefore, for reasons of parsimony and to satisfy computational restrictions, we will analyse the results in Table 7.6 using an exploratory approach.

Table 7.6 suggests the existence of five isolated nodes that can be deemed independent of the others: milk, biscuits, water, coffee and yoghurt. We will therefore try to fit a graphical log-linear model to the remaining 15 variables, in order to see whether the results from the exploratory analysis can be confirmed. Table 7.7 presents the maximum likelihood estimates of the parameters of the log-linear model with interactions up to order 2, fitted on the 15-way contingency table corresponding to the 15 variables considered.

From Table 7.7 it turns out that all interactions found in Table 7.5 remain significant. The difference with Table 7.5 is that we have now taken into account conditional dependences between the variables, and as all variables are positively

Table 7.7   Maximum likelihood estimates of the log-linear parameters.

Parameter	Estimate	Standard error	Chi-square	P-value
TIN_MEAT	−1.6186	0.2206	53.85	<.0001
MOZZAR	−0.01	0.132	0.01	**0.9396**
TIN_MEAT*MOZZAR	0.6607	0.066	100.31	<.0001
TUNA	−0.392	0.0635	38.07	<.0001
TIN_MEAT*TUNA	0.3994	0.0344	134.72	<.0001
MOZZAR*TUNA	0.1483	0.029	26.17	<.0001
COKE	−0.174	0.075	5.38	0.0203
TIN_MEAT*COKE	0.2215	0.0501	19.58	<.0001
MOZZAR*COKE	0.0769	0.0326	5.58	0.0182
TUNA*COKE	0.0592	0.0117	25.63	<.0001
CRACKERS	−0.2079	0.1228	2.87	**0.0904**
TIN_MEAT*CRACKERS	0.4715	0.0768	37.71	<.0001
MOZZAR*CRACKERS	0.1389	0.0616	5.08	0.0242
TUNA*CRACKERS	0.1504	0.0188	63.8	<.0001
COKE*CRACKERS	0.1068	0.0199	28.68	<.0001
PASTA	−0.2935	0.0516	32.35	<.0001
TIN_MEAT*PASTA	0.0294	0.0346	0.72	**0.3957**
MOZZAR*PASTA	0.00751	0.0206	0.13	**0.7156**
TUNA*PASTA	0.0872	0.00796	120.2	<.0001
COKE*PASTA	0.0267	0.00805	11.01	0.0009
CRACKERS*PASTA	0.0219	0.0144	2.3	**0.1291**
JUICES	−0.3191	0.0807	15.62	<.0001
TIN_MEAT*JUICES	0.2942	0.0543	29.32	<.0001
MOZZAR*JUICES	0.1089	0.0347	9.84	0.0017
TUNA*JUICES	0.0879	0.0126	48.95	<.0001
COKE*JUICES	0.1238	0.0119	107.57	<.0001
CRACKERS*JUICES	0.1683	0.02	70.84	<.0001
PASTA*JUICES	0.0304	0.00901	11.41	0.0007
OIL	0.0318	0.1195	0.07	**0.7902**
TIN_MEAT*OIL	0.4508	0.077	34.28	<.0001
MOZZAR*OIL	0.1343	0.0569	5.57	0.0183
TUNA*OIL	0.1219	0.018	45.9	<.0001
COKE*OIL	0.0466	0.0204	5.2	0.0226
CRACKERS*OIL	0.1644	0.0361	20.76	<.0001
PASTA*OIL	0.0792	0.0131	36.54	<.0001
JUICES*OIL	0.063	0.023	7.52	0.0061
TOMATO_J	−0.1715	0.0712	5.8	0.016
TIN_MEAT*TOMATO_J	0.2314	0.0469	24.34	<.0001
MOZZAR*TOMATO_J	0.1121	0.0284	15.62	<.0001
TUNA*TOMATO_J	0.0605	0.0112	29.23	<.0001
COKE*TOMATO_J	0.0958	0.0108	78.92	<.0001
CRACKERS*TOMATO_J	0.0589	0.0211	7.77	0.0053
PASTA*TOMATO_J	0.1887	0.00747	637.43	<.0001
JUICES*TOMATO_J	0.0831	0.0122	46.54	<.0001
OIL*TOMATO_J	0.178	0.0163	119.22	<.0001

Table 7.7 (*continued*)

Parameter	Estimate	Standard error	Chi-square	P-value
BRIOCHES	−0.2412	0.062	15.14	0.0001
TIN_MEAT*BRIOCHES	0.153	0.0414	13.64	0.0002
MOZZAR*BRIOCHES	0.0955	0.0254	14.17	0.0002
TUNA*BRIOCHES	0.0774	0.00995	60.57	<.0001
COKE*BRIOCHES	0.0965	0.00966	99.71	<.0001
CRACKERS*BRIOCHES	0.186	0.0156	141.6	<.0001
PASTA*BRIOCHES	0.0343	0.00689	24.81	<.0001
JUICES*BRIOCHES	0.2342	0.00962	592.76	<.0001
OIL*BRIOCHES	0.0251	0.0176	2.02	**0.1552**
TOMATO_J*BRIOCHES	0.0608	0.00967	39.54	<.0001
BEER	−0.0287	0.0742	0.15	**0.6987**
TIN_MEAT*BEER	0.2098	0.0462	20.62	<.0001
MOZZAR*BEER	0.07	0.0333	4.42	0.0356
TUNA*BEER	0.0864	0.0113	58.89	<.0001
COKE*BEER	0.2415	0.00965	626.64	<.0001
CRACKERS*BEER	0.0721	0.021	11.76	0.0006
PASTA*BEER	0.00755	0.00802	0.89	**0.3464**
JUICES*BEER	0.1201	0.0119	102.14	<.0001
OIL*BEER	0.0805	0.0192	17.61	<.0001
TOMATO_J*BEER	0.0602	0.0111	29.56	<.0001
BRIOCHES*BEER	0.0621	0.00985	39.83	<.0001
FROZ_VEG	−0.2247	0.0704	10.18	0.0014
TIN_MEAT*FROZ_VEG	0.1938	0.049	15.64	<.0001
MOZZAR*FROZ_VEG	0.1211	0.0276	19.25	<.0001
TUNA*FROZ_VEG	0.0634	0.0114	31.14	<.0001
COKE*FROZ_VEG	0.0398	0.0116	11.75	0.0006
CRACKERS*FROZ_VEG	0.063	0.0214	8.7	0.0032
PASTA*FROZ_VEG	0.0381	0.00773	24.3	<.0001
JUICES*FROZ_VEG	0.0496	0.0129	14.76	0.0001
OIL*FROZ_VEG	0.072	0.0188	14.59	0.0001
TOMATO_J*FROZ_VEG	0.0847	0.0106	63.29	<.0001
BRIOCHES*FROZ_VEG	0.0406	0.00993	16.75	<.0001
BEER*FROZ_VEG	0.0224	0.0118	3.61	**0.0575**
RICE	−0.2743	0.084	10.67	0.0011
TIN_MEAT*RICE	0.2987	0.0514	33.83	<.0001
MOZZAR*RICE	0.1887	0.0355	28.34	<.0001
TUNA*RICE	0.146	0.0131	124.67	<.0001
COKE*RICE	0.0626	0.0149	17.65	<.0001
CRACKERS*RICE	0.1909	0.0235	65.96	<.0001
PASTA*RICE	0.1481	0.00975	231.01	<.0001
JUICES*RICE	0.1024	0.0155	43.4	<.0001
OIL*RICE	0.1225	0.0237	26.75	<.0001
TOMATO_J*RICE	0.109	0.0129	70.9	<.0001

(*continued overleaf*)

Table 7.7   (*continued*)

Parameter	Estimate	Standard error	Chi-square	P-value
BRIOCHES*RICE	0.0228	0.0128	3.15	**0.0759**
BEER*RICE	0.0362	0.0151	5.74	0.0166
FROZ_VEG*RICE	0.0949	0.0136	49.02	<.0001
F_FISH	−0.0494	0.1337	0.14	**0.7119**
TIN_MEAT*F_FISH	0.4792	0.0894	28.74	<.0001
MOZZAR*F_FISH	0.2417	0.0527	21.04	<.0001
TUNA*F_FISH	0.1034	0.0224	21.4	<.0001
COKE*F_FISH	0.0504	0.0258	3.82	**0.0507**
CRACKERS*F_FISH	0.1047	0.0494	4.48	0.0342
PASTA*F_FISH	0.0536	0.0156	11.78	0.0006
JUICES*F_FISH	0.1032	0.0274	14.22	0.0002
OIL*F_FISH	0.1232	0.0419	8.66	0.0033
TOMATO_J*F_FISH	0.075	0.0221	11.49	0.0007
BRIOCHES*F_FISH	0.0545	0.0207	6.92	0.0085
BEER*F_FISH	0.0735	0.0243	9.16	0.0025
FROZ_VEG*F_FISH	0.2954	0.0169	305.75	<.0001
RICE*F_FISH	0.1711	0.0262	42.64	<.0001
ICECREAM	−0.4074	0.1882	4.68	0.0304
TIN_MEAT*ICECREAM	0.6214	0.1579	15.49	<.0001
MOZZAR*ICECREAM	0.1597	0.0828	3.73	**0.0536**
TUNA*ICECREAM	0.1106	0.0293	14.28	0.0002
COKE*ICECREAM	0.2095	0.0235	79.35	<.0001
CRACKERS*ICECREAM	0.2912	0.0417	48.72	<.0001
PASTA*ICECREAM	−0.00983	0.02	0.24	**0.6233**
JUICES*ICECREAM	0.2335	0.0255	83.69	<.0001
OIL*ICECREAM	0.1632	0.0534	9.33	0.0023
TOMATO_J*ICECREAM	0.0961	0.0286	11.31	0.0008
BRIOCHES*ICECREAM	0.1393	0.022	40.05	<.0001
BEER*ICECREAM	0.1133	0.0278	16.57	<.0001
FROZ_VEG*ICECREAM	0.2202	0.024	84.07	<.0001
RICE*ICECREAM	0.1967	0.0347	32.13	<.0001
F_FISH*ICECREAM	0.1872	0.056	11.18	0.0008

associated, more interactions have been found significant. Table 7.7 reveals no significant negative interactions.

7.4.2   Association rules

The most common way to analyse market basket data is to use association rules, a local data mining method explained in Section 4.8. We begin with a simple setting. Consider the products ice cream and Coke. As order is not relevant, to study the association between the two products, the data set can be collapsed to the two-way contingency table in Table 7.4. This shows that the support for the rule 'if ice cream, then Coke' is

$$\text{support(ice cream} \rightarrow \text{Coke)} = \frac{170}{46\,727} = 0.0036,$$

indicating low support for the rule. This means these two products are bought together only occasionally. The support corresponds to only one of the four joint frequencies in Table 7.4, corresponding to the occurrence of both buying events. A support of 0.0036 means that only 0.36% of the transactions considered will have both ice cream and Coke in the basket. The support of an association rule is symmetric; the support of the rule 'if Coke, then ice cream' would be the same.

The confidence of a rule, even when calculated for an association, where order does not matter, depends on the body and head of the rule:

$$\text{confidence(ice cream} \rightarrow \text{Coke)} = \frac{170}{769} = 0.22,$$

which corresponds to the second row conditional frequency of Coke $= 1$, and

$$\text{confidence(Coke} \rightarrow \text{ice cream)} = \frac{170}{4949} = 0.034,$$

which corresponds to the second column conditional frequency of ice cream $= 1$. The confidence is really a particular conditional frequency. In the first case it indicates the proportion, among those who buy ice cream, of those who also buy Coke. In the second case it indicates the proportion, among those who buy Coke, of those who also buy ice cream. The lift is a normalised measure of interestingness; it is also symmetric:

$$\text{lift(ice cream} \rightarrow \text{Coke)} = \frac{0.22}{0.11} = 2, \text{lift(Coke} \rightarrow \text{ice cream)} = \frac{0.034}{0.017} = 2.$$

This is always the case, as can be seen from the formula in Section 4.8. Section 4.8 goes on to derive an asymptotic confidence interval for the lift. Here the asymptotic confidence interval goes from 1.17 to 3.40, so the association can be considered significant.

Notice that the odds ratio between the two products was calculated as 2.44, a rather similar value (and also with a significant confidence interval). The main difference is that the odds ratio depends explicitly on all four cell frequencies of a contingency table, whereas the lift is the ratio between the frequency of the levels $(A = 1, B = 1)$ and the product of the two marginal frequencies, $(A = 1)$ and $(B = 1)$, so it depends only implicitly on the frequencies of the complementary events $(A = 0, B = 0)$.

In any case the support of the rule considered is rather limited, – ice cream and Coke are present in only 0.36% of all transactions – therefore conclusions based on it may not be of much practical value, even when supported by a high confidence and/or a high lift value. But this conclusion is relative; it depends on the support of other rules. To discover this and obtain a more comprehensive picture of the interesting association rules, we now move to a full application of association rule modelling. The Apriori algorithm and a threshold support rule of 0.05*support(mode), where mode is the rule with maximum support among all rules of a fixed order, leads to the selection of several relevant rules.

Table 7.8 lists the most frequent transactions of order 2, giving their order, frequency, and the two products involved in each transaction. Tables 7.9 and 7.10 list the transactions of order 3 and 4.

Table 7.11 presents the order 2 association rules with highest support. It shows, for example, that pasta → milk has support equal to 49.84. Table 7.12 presents the order 4 association rules with highest confidence, with the same notation as in Table 7.11. We can see, for example, that tuna – tomato sauce – crackers → pasta has a confidence equal to 98.95. This means that, if a transaction contains pasta, it will also contain tuna, tomato sauce and crackers about 99% of the time. On the other hand, the converse rule pasta → tuna – tomato sauce – crackers is not among those with the highest confidence. Indeed, it can be shown to have a confidence equal to 22.78. The latter can be interpreted saying that, if a transaction contains pasta, it will also contain tuna, tomato sauce and crackers only about 23% of the time.

Next we try a methodology based on tree models (Section 4.8). We have chosen pasta, the most frequent product and the most frequent head of the rule in the associations. We have constructed a tree model having pasta as target variable and all the other products as predictors. Among the different paths leading to the terminal nodes, we consider those paths where all variables in the path have the value 1. These paths corresponds to rules with high confidence. Using a CHAID tree (CART gives similar results), we obtain the following rules:

tuna – tomato sauce →pasta
tomato sauce – rice →pasta
rice – biscuits →pasta

and their respective measures of interestingness:

lift 1.41, confidence 95.24%, support 14.84%
lift1.44, confidence 96.80%, support 12.14%
lift1.40, confidence 94.23%, support 18.43%

Notice that all three rules have high confidence. This is to be expected, as a tree model tries to develop the best predictive rules for the target variable.

7.5   Model comparison

It is quite difficult to assess local models such as association rules, since model evaluation measures apply to global models. Furthermore, as the idea of searching for local patterns and rules is very recent, there is little consensus in the data mining literature on how to measure their performance (Hand *et al.*, 2001). A natural idea is to measure the utility of patterns in terms of how interesting or unexpected they are to the analyst. As it is quite difficult to model an analyst's opinion, we usually assume a situation of completely uninformed opinion. In this

Table 7.8 Transaction count of order 2.

Order	Frequency	Product 1	Product 2
2	3359	pasta	milk
2	2686	milk	biscuits
2	2677	water	pasta
2	2675	water	milk
2	2238	pasta	coffee
2	2204	milk	coffee
2	2154	pasta	brioches
2	2146	water	biscuits
2	2095	milk	brioches
2	2084	tuna	pasta
2	2084	pasta	frozen veg
2	2003	yoghurt	milk
2	1993	milk	frozen veg
2	1971	yoghurt	pasta
2	1943	tuna	milk
2	1825	coffee	biscuits
2	1821	brioches	biscuits
2	1807	water	coffee
2	1743	tomato sauce	pasta
2	1735	water	brioches
2	1679	water	frozen veg
2	1655	yoghurt	biscuits
2	1650	tuna	biscuits
2	1643	rice	pasta
2	1636	frozen veg	biscuits
2	1621	water	tuna
2	1604	tomato sauce	milk
2	1602	yoghurt	water
2	1595	pasta	beer
2	1558	milk	beer
2	1557	pasta	coke
2	1512	milk	coke
2	1503	rice	milk
2	1441	coffee	brioches
2	1427	water	beer
2	1421	tuna	coffee
2	1398	pasta	juices
2	1387	milk	juices
2	1368	frozen veg	coffee

Table 7.9   Transaction count of order 3.

ORDER	Frequency	Product 1	Product 2	Product 3
3	2287	pasta	milk	biscuits
3	2258	water	pasta	milk
3	1926	pasta	milk	coffee
3	1888	water	milk	biscuits
3	1872	water	pasta	biscuits
3	1810	pasta	milk	brioches
3	1768	pasta	milk	frozen veg
3	1736	tuna	pasta	milk
3	1721	yoghurt	pasta	milk
3	1633	pasta	coffee	biscuits
3	1627	milk	coffee	biscuits
3	1600	water	pasta	coffee
3	1599	water	milk	coffee
3	1589	pasta	brioches	biscuits
3	1573	milk	brioches	biscuits
3	1523	water	pasta	brioches
3	1507	tuna	pasta	biscuits
3	1499	water	milk	brioches
3	1495	pasta	frozen veg	biscuits
3	1494	water	pasta	frozen veg
3	1480	tomato sauce	pasta	milk
3	1479	water	milk	frozen veg
3	1476	yoghurt	milk	biscuits
3	1468	water	tuna	pasta
3	1468	milk	frozen veg	biscuits
3	1457	yoghurt	pasta	biscuits
3	1455	tuna	milk	biscuits
3	1418	yoghurt	water	milk
3	1412	water	tuna	milk
3	1405	yoghurt	water	pasta

case study we have considered support, confidence and lift as the main measures for validating a set of association rules. But the needs of the user will govern which of these three is the best one for selecting a set of rules. The support can be used to assess the importance of a rule in terms of its frequency in the database; the confidence can be used to investigate possible dependences between variables; and the lift can be used to measure the distance from the situation of independence.

Ultimately, a set of rules has to be assessed on its ability to meet the analysis objectives. Here the objectives are primarily to reorganise the layout of a sales outlet and to plan promotions so as to increase revenues. Once the associations have been identified, it is possible to organise promotions within the outlet so

Table 7.10 Transaction count of order 4.

Order	Frequency	Product 1	Product 2	Product 3	Product 4
4	1693	water	pasta	milk	biscuits
4	1488	pasta	milk	coffee	biscuits
4	1455	water	pasta	milk	coffee
4	1412	pasta	milk	brioches	biscuits
4	1363	water	pasta	milk	brioches
4	1358	pasta	milk	frozen veg	biscuits
4	1357	tuna	pasta	milk	biscuits
4	1352	water	pasta	milk	frozen veg
4	1336	yoghurt	pasta	milk	biscuits
4	1312	water	tuna	pasta	milk
4	1286	yoghurt	water	pasta	milk
4	1259	water	milk	coffee	biscuits
4	1255	water	pasta	coffee	biscuits
4	1209	water	pasta	brioches	biscuits
4	1205	water	milk	brioches	biscuits
4	1174	tuna	pasta	milk	coffee
4	1174	pasta	milk	coffee	brioches
4	1172	water	pasta	frozen veg	biscuits
4	1165	water	milk	frozen veg	biscuits
4	1155	pasta	milk	frozen veg	coffee
4	1143	water	tuna	pasta	biscuits
4	1131	yoghurt	water	milk	biscuits
4	1122	water	tuna	milk	biscuits
4	1121	rice	pasta	milk	biscuits
4	1115	yoghurt	water	pasta	biscuits
4	1112	tomato sauce	pasta	milk	biscuits
4	1103	yoghurt	pasta	milk	coffee
4	1100	tuna	pasta	milk	frozen veg
4	1099	water	pasta	milk	beer
4	1098	pasta	milk	frozen veg	brioches
4	1091	water	tomato sauce	pasta	milk
4	1090	tuna	pasta	milk	brioches
4	1060	yoghurt	pasta	milk	brioches
4	1056	water	rice	pasta	milk
4	1051	water	pasta	milk	coke
4	1051	pasta	milk	coke	biscuits
4	1051	pasta	milk	biscuits	beer
4	1045	tuna	pasta	coffee	biscuits
4	1045	pasta	coffee	brioches	biscuits
4	1040	yoghurt	pasta	milk	frozen veg
4	1038	milk	coffee	brioches	biscuits
4	1022	tuna	milk	coffee	biscuits
4	1020	water	tuna	pasta	coffee
4	1015	yoghurt	tuna	pasta	milk
4	1011	pasta	frozen veg	coffee	biscuits
4	1007	water	pasta	coffee	brioches
4	1003	milk	frozen veg	coffee	biscuits
4	1002	water	pasta	frozen veg	coffee
4	1001	water	milk	frozen veg	coffee

Table 7.11 Transaction count ordered by support (S); EC = expected confidence, C = confidence, L = lift, F = frequency.

EC	C	S	L	F	Product 1	Product 2
65.70	73.97	49.84	1.12	3359	pasta	milk
67.38	75.85	49.84	1.12	3359	milk	pasta
50.02	59.21	39.90	1.18	2689	pasta	biscuits
67.38	79.76	39.90	1.18	2689	biscuits	pasta
50.02	60.65	39.85	1.21	2686	milk	biscuits
65.70	79.67	39.85	1.21	2686	biscuits	milk
67.38	77.54	39.72	1.15	2677	water	pasta
51.22	58.95	39.72	1.15	2677	pasta	water
65.70	77.49	39.69	1.17	2675	water	milk
51.22	60.41	39.69	1.17	2675	milk	water

Table 7.12 Transaction count sorted by Confidence (C).

EC	C	S	L	F	Product 1	Product 2
67.38	98.95	4.21	1.46	284	tuna – tomato sauce – crackers	pasta
67.38	98.80	3.68	1.46	248	tomato sauce – crackers – coke	pasta
67.38	98.67	6.61	1.46	446	yoghurt – rice – juices	pasta
67.38	98.51	5.89	1.46	397	tomato sauce – rice – juices	pasta
67.38	98.49	3.87	1.46	261	tuna – oil – brioches	pasta
67.38	98.43	3.72	1.46	251	tuna – rice – oil	pasta
67.38	98.39	3.63	1.46	245	rice – oil – frozen veg	pasta
67.38	98.36	3.57	1.45	241	frozen fish – coke – biscuits	pasta
67.38	98.35	3.54	1.45	239	tomato sauce – rice – crackers	pasta
67.38	98.33	8.75	1.45	590	tomato sauce – rice – frozen veg	pasta
67.38	98.33	3.50	1.45	236	rice – juices – crackers	pasta
67.38	98.32	8.72	1.45	588	tomato sauce – rice – coffee	pasta

that the products that are put on offer at the same time are products which are not associated. Correspondingly, by putting one product on promotion, we also increase the sales of the associated products.

At the beginning of the chapter we saw that odds ratios and log-linear models can also be employed to determine a global association structure between the buying variables; in this case traditional statistical measures, such as G^2, or AIC and BIC, can be employed to assess the overall quality of a model. Although they have a different purpose, classification trees can also be seen as a global model capable of producing an association structure. Although association rules are much easier to detect and interpret, good global modelling, as expressed by log-linear and tree models, allows more stable and coherent conclusions. With enough time and sufficient knowledge to implement a global model, this approach should be preferred.

7.6 Summary report

1. **Context.** This case study concerns the understanding of associations between buying behaviours. A similar kind of analysis can be applied to problems in which the main objective is cross-selling to increase the number of products that are sold in a given commercial unit (a supermarket, a bank, a travel agency, or, more generally, a company offering more than one product or service). A related class of problems arise in promotional campaigns: it is desirable to put on promotion the smallest possible number of products but to derive any benefits on the largest possible number of products. This is achieved by an efficient layout of the products, putting together those that are most associated with each other.

2. **Objectives.** The aim of the analysis is to track the most important buying patterns, where a pattern means a group of products bought together. The most common measures refer either to the probability of buying a certain basket of products (support) or to the conditional probability of buying a certain product, having bought others (confidence).

3. **Organisation of the data.** Data is extracted from a large database containing all commercial transactions in a supermarket in a given amount of time. The transactions are made by someone holding one of the chain's loyalty cards. Although data is structured in a transactional database, it can be simplified into a data matrix format, with rows identifying clients and columns associated with binary variables describing whether or not each product has been bought. After some simplification the data matrix contains 46 727 rows and 20 columns.

4. **Exploratory data analysis.** Exploratory data analysis was performed by looking at all pairwise odds ratios between the 20 products, for a total of 190 association measures. The results can be graphically visualised and already give important suggestions for the objectives of the analysis.

5. **Model specification.** The data mining method employed here is a local model. We compared association rules, originally devised for market basket analysis problems, with more structured log-linear models, the most important symmetric statistical models for the analysis of contingency tables.

6. **Model comparison.** It is rather difficult to compare association rules, which are local, with log-linear models, which are global. The most effective measure of comparison has to be the practical utility of a rule; this can be measured in terms of cross-selling or by using the efficacy of a promotional campaign.

7. **Model interpretation.** Association rules seem to be easier to understand than the results from a log-linear models, but it depends on how the results are presented. Results from a log-linear model can be expressed graphically, using dependences and odds ratios, and these measures are easy to understand. The advantage of log-linear models is that they are based on inferential statements and can therefore provide confidence intervals for an association statement or a threshold able to 'filter' out relevant rules from the many possible rules and in a coherent way.

CHAPTER 8

Describing customer satisfaction

8.1 Objectives of the analysis

This chapter is concerned with data mining methods for customer satisfaction analysis. Customer satisfaction is a measure of how the products and services supplied by a company match customer expectations. To enable it to be measured statistically, customer satisfaction must be translated into a number of measurable indicators, directly linked to factors that can be understood and influenced. For more details, see Siskos *et al.* (1998), Cassel (2000), Cassel and Eklöf (2001), Cassel *et al.* (2002) and Särndal and Lundström(2005).

Satisfaction is a somewhat vague concept, but it can be measured by simply asking a series of questions. A customer may be completely satisfied with the quality of a service, not satisfied at all, or somewhere in between. We can take 'not satisfied at all' and 'completely satisfied' as fixed endpoints of an ordinal variable and then we have to decide how many points there should be in between. It would be ideal if satisfaction could be measured on a continuous scale. But for obvious reasons this is impossible and we have to compromise. The scale should be such that it allows the customer enough flexibility to accurately express an opinion. For example, the customer may be asked to indicate which of the following best describes his or her views: 1, very unsatisfied; 2, moderately unsatisfied; 3, neutral; 4, moderately satisfied; 5, very satisfied. Questions presented in this way are scored on a five-point scale. Overall a questionnaire will contain some 30–40 questions about the customer's satisfaction with different aspects of the service. There should also be some background variables on the customer that will make it possible to do a more detailed analysis.

To estimate customer satisfaction descriptive statistical methods are typically used. The recent literature suggests comparing statistical models for customer satisfaction in terms of predictive performance. Guidelines provided by international quality organizations such as the European Foundation for Quality Management (EFQM), the European Quality Organisation (EOQ) and national quality organisations suggests using structures of latent variables.

Applied Data Mining for Business and Industry, 2e   P. Giudici, S. Figini
© 2009 John Wiley & Sons, Ltd

8.2   Description of the data

The data set analysed in this chapter comes from the ABC 2004 annual customer satisfaction survey. The survey was carried out by KPA Ltd., an independent consulting firm partner of the European Musing project (www.musing.eu). ABC, a software house (whose name has been changed to protect its identity), wished to measure customer satisfaction on the part of its customers. It collected information on:

- overall satisfaction levels;
- equipment (e.g. 'Improvements and upgrades provide value');
- sales support (e.g. 'Sales personnel respond promptly to requests');
- technical support (e.g. 'Technical support is available when needed' and 'The technical staff is well informed about the latest equipment updates/enhancements');
- training (e.g. 'The trainers are knowledgeable about the equipment' and 'The trainers are effective communicators');
- supplies and media (e.g. 'ABC branded performance meets your expectations');
- pre-press/workflow and post-press solutions (e.g. 'Capabilities and features of tools meet your needs');
- customer portal (My ABC) (e.g. 'The portal's resources are helpful');
- administrative support;
- terms, conditions, and pricing (e.g. 'Equipment and service contract terms are clear');
- site planning and installation (e.g. Equipment worked properly after installation);
- overall satisfaction with competitors.

There were 81 questions in total; in most cases the level of satisfaction is measured on a five-point scale from very low satisfaction (1) to very high satisfaction (5). The qualitative variables derived are thus qualitative and ordinal. A total of 261 customers eventually took part in the questionnaire.

8.3   Exploratory data analysis

Most of the items in the questionnaire take the form of a statement describing the customer's experience with ABC during 2004. The person filling in the questionnaire also gives his or her title or position, the company's geographical location, and the length in years of its relationship with ABC.

The first part of the questionnaire deals with 'Overall satisfaction'. There are four questions to be assessed on a five-point scale, and one to be answered 'yes' or 'no' which asks whether ABC is the respondent's best supplier. Two of the questions imply a comparison with other companies, one asking whether the customer would buy a given product from ABC rather than someone else, asking whether the customer would recommend ABC to other companies; these are scored on a five-point scale from very unlikely to very likely.

Then there are blocks of questions on equipment, sales support, technical support, training, supplies and media, pre-press/workflow and post-press solutions, customer portal, administrative support, terms, conditions, and pricing, site planning and installation, and overall satisfaction with other ABC's competitors. The customer marks his or her level of agreement with statement on a five-point scale from 1 (strongly disagree) to 5 (strongly agree), and then assesses the level of importance of the statement on a three-point scale (1, low; 2, medium; or 3, high). Any statement that is not relevant or not applicable can be marked N/A.

After a descriptive data analysis phase during which Questions 68–81 (on overall satisfaction with competitors) are deleted, we have a data set consisting of 67 variables on 240 customers. We give a short summary of the exploratory analysis.

Concerning overall satisfaction with ABC, only 91 customers consider ABC their best supplier. The results of Question 11, which measures overall satisfaction with the equipment, are shown in Figure 8.1: 54% of the customers report high satisfaction and 28% medium level of satisfaction. Figure 8.2 shows the overall satisfaction with sales support (Question 17): only a few customers (33) are very highly satisfied. On the other hand, as far as technical support is concerned, 99 customers are highly satisfied and 68 very highly satisfied. The overall satisfaction with ABC's supplies and media is medium, and with workflow solutions very high. There is a high overall satisfaction with the administrative support and a medium level of satisfaction with terms, conditions and pricing. Regarding overall satisfaction with overall solutions with the customer portal, site planning and installation, much of the data is missing (the number of non-responses is very high).

Statistics on customer seniority and country location are reported in Figures 8.3 and 8.4. Germany accounts for 44% of the customers. Only a small percentage of the customers are located in France and Israel. In terms of customer seniority (i.e. the length of the relationship between ABC and the customer), we observe a quite high percentage of old and new customers, but a dip in the number of

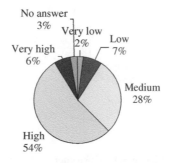

Figure 8.1 Overall satisfaction with equipment (Question 11).

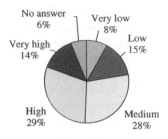

Figure 8.2   Overall satisfaction with sales support (Question 17).

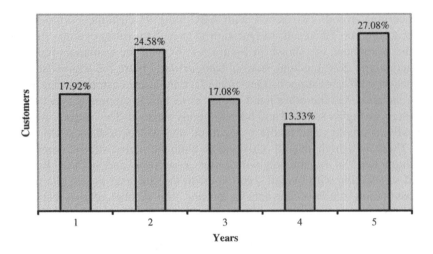

Figure 8.3   Customer seniority.

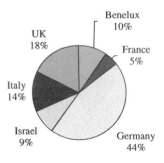

Figure 8.4   Customer location.

companies which have been customers for 4 years (13%). This suggests that it could be interesting to examine how customer satisfaction changes over the years. We can extend the descriptive analysis to measure possible associations among the variables. However, we prefer to report here directly the results based on discrete graphical models. As our target variable we use overall satisfaction a binary variable that reports whether a customer is satisfied ($Y = 0$) or not satisfied ($Y = 1$).

8.4 Model building

We analyse our data using discrete graphical models, a class of models that can be used both descriptively and predictively (see section 4.14) for categorical data. In graphical models the nodes represent variables, and the edges between nodes represent conditional dependences between them. Here we consider models that are used to represent associations between a number of discrete variables. We start with the so-called log-linear representation of a multi-way contingency table. This is convenient for our purposes because it allows us to express (conditional) independence constraints by setting certain coefficients equal to zero.

We consider the problem of finding a good model when little or no prior knowledge is available on the independence/dependence relationships among the variables. We discuss two approaches to model selection: one based on significance testing and one based on a model quality criterion. In both cases we use stepwise selection, which is an incremental search procedure. Starting from an initial model, edges are successively added or removed until some criterion is achieved (see e.g. Edwards, 2000). At each step the inclusion or exclusion of eligible edges is decided using significance tests. Eligible edges are tested for removal using chi-squared tests based on the difference in deviance between successive models. The edge whose chi-squared test has the largest non-significant p-value is removed. If all p-values are significant (i.e., all $p < \alpha$, where α is the significance level), then no edges are removed and the procedure stops. In our application we also use stepwise model selection using Akaike's information criterion, which assigns to any model M the measure $\text{AIC}(M) = \text{dev}(M) + 2$, where p is the number of parameters in the model. This quality measure consists of two components: the lack of fit of the model as measured by the deviance, and the complexity of the model as measured by the number of parameters. We remove those edges that provide the largest reduction in AIC. The software used for this procedure defines the AIC somewhat differently, namely $\text{AIC}(M) = -2L^M + 2p$, where L^M is the value of the log-likelihood function evaluated at $\hat{p}^M$, the ML estimate of p under M. It is easy to see that this formulation is the same as the previous one, since $\text{dev}(M) = 2(L^{\text{sat}} - L^M) = 2L^{\text{sat}} - 2L^M$, where L^{sat} is the value of the log-likelihood function of the saturated model evaluated at its maximum.

Concerning model specification, we analyse different sets of variables. First we consider as target variable the overall satisfaction level with ABC, labelled a. To explain this variable we use the following questions:

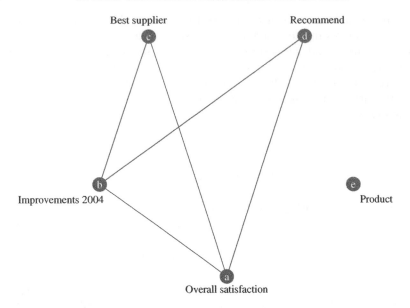

Figure 8.5   The first selected graphical model.

- Overall satisfaction level with ABC's improvements during 2004 (Question 2, labelled *b*).
- Is ABC your best supplier? (Question 3, labelled *c*).
- Would you recommend ABC to other companies? (Question 4, labelled *d*).
- If you were in the market to buy a product, how likely would it be for you to purchase an ABC product again? (Question 5, labelled *e*).

Figure 8.5 shows that Question 5 is not related to the other variables. We observe a relationship among the other variables and, in particular, among the questions labelled *abc* and *abd*; finally, *c* and *d* are conditionally independent variables. We have obtained this graphical model in two ways. First we build a model based on backward selection. We compare the model obtained with that obtained by minimisation of AIC, based on stepwise selection. The results for the first procedure whose graph is given in Figure 8.5, show that the model based on stepwise selection is preferable, as it has the smaller AIC.

Retaining our target variable of overall satisfaction level with ABC, labelled *a*, we now take a different set of covariates:

- Overall satisfaction level with the equipment (Question 11, labelled *b*).
- Overall satisfaction level with sales support (Question 17, labelled *c*).
- Overall satisfaction with technical support (Question 25, labelled *d*).
- Overall satisfaction level with ABC training (Question 31, labelled *e*).
- Overall satisfaction level with ABC's supplies and media (Question 38, labelled *f*).

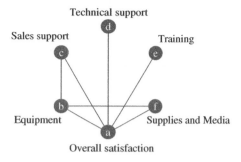

Figure 8.6 The second selected graphical model.

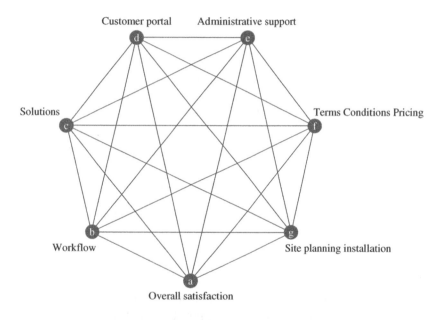

Figure 8.7 The third selected graphical model.

Figure 8.6 shows a number of two-way dependences, between *ac, ad, ae, af, bf, ab*, for model 2. Our result for this model show that the model based on stepwise selection is preferable, as it has the smallest AIC.

Figures 8.7 and 8.8 present two more models based on stepwise selection, with the same target variable *a*. In Figure 8.7 the covariates are:

- Overall satisfaction level with workflow solutions (Question 42, labelled *b*).
- Overall satisfaction with overall solutions for various problems (Question 43, labelled *c*).
- Overall satisfaction level with the customer portal (Question 49, labelled *d*).

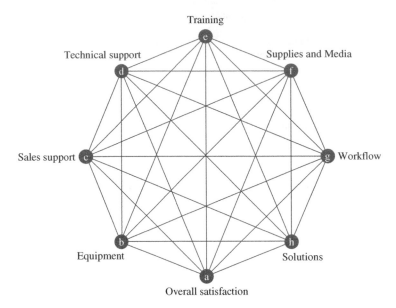

Figure 8.8   The fourth selected graphical model.

- Overall satisfaction level with administrative support (57, e Question 57, labelled *e*).
- Overall satisfaction with terms, conditions, and pricing (Question 65, labelled *f*).
- Overall satisfaction level with site planning and installation (Question 67, labelled *g*).

In Figure 8.8 they are:

- Overall satisfaction level with the equipment (Question 11, labelled *b*).
- Overall satisfaction level with sales support (Question 17, labelled *c*).
- Overall satisfaction with technical support (Question 25, labelled *d*).
- Overall satisfaction level with ABC training (Question 31, labelled *e*).
- Overall satisfaction level with ABC's supplies and media (Question 38, labelled *f*).
- Overall satisfaction level with workflow solutions (Question 42, labelled *g*).
- Overall satisfaction with overall solutions for various problems (Question 43, labelled *h*).

Both figures 8.7 and 8.8 show a highly interdependent association structure.

Finally, we model Overall satisfaction level with ABC (Question 1, labelled b) as a function of :

- Is ABC your best supplier (Question 3, labelled a)

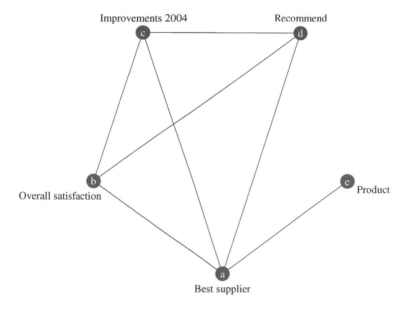

Figure 8.9 The fifth selected graphical model.

- Overall satisfaction level with ABC's improvements during 2004 (Question 2, labelled *c*).
- Would you recommend ABC to other companies? (Question 4, labelled *d*).
- If you were in the market to buy a product, how likely would it be for you to purchase an ABC product again (Question 5, labelled *e*).

To obtain the model in Figure 8.9, which is less connected than the previous two, we have used backward selection, which has the smaller AIC.

We finally remark that all obtained graphical models are descriptive; however, they can be easily be made predictive by choosing one variable as target response, for example the overall satisfaction.

8.5 Summary

1. **Context.** Marketing and management sciences are concerned with the coordination of all the organisation's activities in order to provide goods or services that best satisfy the specific needs of existing or potential customers. Customer satisfaction measures offer meaningful and objective feedback about client preferences and expectations.
2. **Objectives.** Keeping customers satisfied is the only way to stay competitive in today's marketplace. Customers have expectations of service and product performance that must be met. A balancing act between what customers want and what a company can provide must be performed in order to maximise

the firm's long-term profits. With precise information, as derived from surveys, companies can focus on issues that truly drive customer satisfaction. A directed focus often leads to cost reductions because companies can emphasise improvements in areas of customer concern. Focusing on motivators of customer satisfaction leads to more loyal customers, who tend to be the most profitable customers (i.e. repeat business is usually the most profitable). An inclusive customer satisfaction and loyalty programme can therefore be considered a source of future profits.

3. **Organisation of the data.** The data analysed comes from a survey. Each row in the data matrix represents a customer and each column a survey question. All the variables are qualitative.

4. **Exploratory data analysis.** The exploratory data analysis is based on a set of descriptive measures for qualitative variables. We presented a set of graphs to clarify this analysis.

5. **Model specification.** In this chapter we have presented a graphical modelling approach for analysng customer satisfaction data. In order to understand the results we suggest investigating the found associations between all discrete variables. We suggest using graphical log-linear models to study interaction between variables.

6. **Model comparison.** The data mining analysis that needs to be employed here is a descriptive model can be easily transformed into a predictive one. We have compared basic models with more structured graphical models, which constitute the most important symmetric statistical models for the analysis of contingency table data.

7. **Model interpretation.** Graphical models are easy to understand. Their results are based on inferential statements, and thus confidence intervals for an association statement can be easily obtained.

CHAPTER 9

Predicting credit risk of small businesses

9.1 Objectives of the analysis

According to the Basel II capital accord, financial institutions require transparent benchmarks of creditworthiness to structure their risk control systems, facilitate risk transfer through structured transactions and comply with regulatory changes. Traditionally, producing accurate credit risk measures has been relatively straightforward for large companies and retail loans, resulting in high levels of transparency and liquidity in the risk transfer market for these asset classes. The task has been much harder for small and medium size enterprises (SMEs).

The causes of default can be seen as consisting of a number of components: a static component, determined by the characteristics of the SME; a dynamic component that includes trends and the contacts of the SME with the bank over different years; a seasonal part, tied to the period of investment; and external factors that include the course of the markets.

The process of credit scoring is very important for banks as they need to discriminate between good and bad SMEs in terms of creditworthiness. This is a classic example of asymmetric information, where a bank has to reveal hidden data about its customers. Seminal contributions on the subject of default prediction are Altman (1968) and Beaver (1966), and more recent ones include Shumway (2001) and Chava and Jarrow (2004). Statistical methods for evaluating default probability estimates are discussed in Sobehart and Keenan (2001), Engelmann *et al.* (2003) and Stein (2005).

9.2 Description of the data

The data considered in this case study is yearly data, from 1996 to 2004, on 1003 firms belonging to 352 different business sectors from one of the major ratings agencies for SMEs in Germany belonging to the European MUSING project (www.musing.eu). The data set consists of a binary response variable, solvency, and a set of explanatory variables given by financial ratios and time variables. In particular, our data set consists of two tables: one of good SMEs and one of bad

Applied Data Mining for Business and Industry, 2e   P. Giudici, S. Figini
© 2009 John Wiley & Sons, Ltd

SMEs. The latter consists of 708 data for 236 companies, the former of 2694 data for 898 companies. The tables show the ID of the firms, the accounting year, the business sector, the solvency status (0 = solvency; 1 = insolvency), and finally the balance sheet information.

Given this understanding of our balance sheet data and how it is constructed, we can discuss some techniques used to analyse the information there contained. The main way this can be done is through financial ratio analysis. This typically uses ratios to gain insight into the company and its operations. Using financial ratios (such as the debt–equity ratio) can give a better idea of the company's financial condition along with its operational efficiency. It is important to note that some ratios will need information from more than one financial statement, such as from the balance sheet and the income statement.

The main types of ratios that use information from the balance sheet are financial strength ratios and activity ratios. Financial strength ratios, such as the debt–equity ratio, provide information on how well the company can meet its obligations and how they are leveraged. This can give investors an idea of how financially stable the company is and how the company finances itself. Activity ratios focus mainly on current accounts to show how well the company manages its operating cycle. These ratios can provide insight into the operational efficiency of the company.

There are a wide range of individual financial ratios that can be calculated to learn more about a company. We computed a set of 11 financial ratios used by subject-matter experts:

- **Supplier target.** This is a temporal measure of financial sustainability expressed in days that considers all short- and medium-term debts as well as other payables.
- **Outside capital structure.** This ratio evaluates the ability of the company to receive other forms of financing beyond banks' loans.
- **Cash ratio.** This indicates the cash a company can generate in relation to its size.
- **Capital tied up.** This ratio evaluates the turnover of short-term debts with respect to sales;
- **Equity ratio.** This is measure of a company's financial leverage calculated by dividing a particular measure of equity by the firm's total assets.
- **Cash flow to effective debt.** This ratio indicates the cash a company can generate in relation to its size and debts.
- **Cost–income ratio.** This is an efficiency measure similar to the operating margin that is useful to measure how costs are changing compared to income.
- **Trade payable ratio.** This reveals how often the firm's payables turn over during the year: a high ratio means a relatively short time between purchase of goods and services and payment for them, while a low ratio may be a sign that the company has chronic cash shortages.
- **Liabilities ratio.** This is a measure of a company's financial leverage calculated by dividing a gross measure of long-term debts by firm assets; it indicates what proportion of debt the company is using to finance its assets.

- **Result ratio.** This is an indicator of how profitable a company is relative to its total assets; it gives an idea as to how efficient management is at using its assets to generate earnings.
- **Liquidity ratio.** This measures the extent to which a firm can quickly liquidate assets and cover short-term liabilities, and therefore is of interest to short-term creditors.

Furthermore, we considered some additional annual account positions, which were standardized in order to avoid computational problems with the previous ratios:

- **Total assets.** This is the sum of current and long-term assets owned by the firm.
- **Total equity.** This refers to total assets minus total liabilities, and it is also referred to as equity or net worth or book value.
- **Total liabilities.** This includes all the current liabilities, long-term debt, and any other miscellaneous liabilities the company may have.
- **Sales.** This is represented by one-year total sales.
- **Net income.** This is equal to the income that a firm has after subtracting costs and expenses from the total revenue.

Finally an important variable is called 'creditworthiness'. The index ranges from 100 to 600 points. A value of 100 means that the company is very creditworthy, whereas 500 means that the company has massive payment problems, and 600 means that the company is insolvent.

9.3 Exploratory data analysis

Based on our data, we have computed for each financial ratio classical tendency measures and variability measures, which we will now briefly summarise.

The response variable, solvency, shows a different degree of incidence during the years considered. In particular, the default events in 1996, 1997, 1998, 1999, 2001, 2002, 2003 and 2004 are respectively: 38, 64, 19, 12, 24, 39, 44, 16 and 4. The information is summarised in Figure 9.1.

To compare the variability of the quantitative variables, we computed the coefficient of variation for the good SMEs (solvency $= 1$) and for the bad SMEs (solvency $= 0$). Some financial ratios show high levels of variability. In particular, as shown in Figure 9.2, for the good SMEs, the cash ratio, cash flow to effective debt, result ratio and liquidity ratio are the most variable. For the bad SMEs, on the other hand, the most variable financial ratios are supplier target days, capital tied up cash flow to effective debt and result ratio.

The average creditworthiness index for the good SMEs is 188.76 and for the bad SMEs is 597.31. To assess the variability of the creditworthiness index, we employ in a different context the ROC curve described in Section 5.5. Based on the ROC curve shown in Figure 9.3, the overall Gini index for the good SMEs

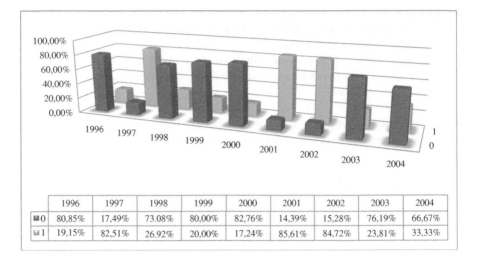

	1996	1997	1998	1999	2000	2001	2002	2003	2004
0	80,85%	17,49%	73.08%	80,00%	82,76%	14,39%	15,28%	76,19%	66,67%
1	19,15%	82,51%	26.92%	20,00%	17,24%	85,61%	84,72%	23,81%	33,33%

Figure 9.1 Calendar time default distribution.

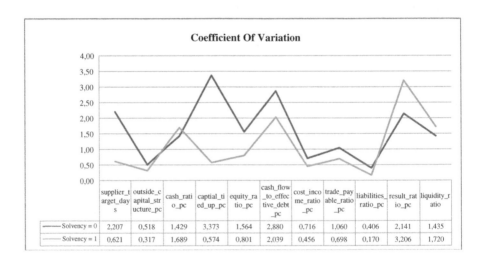

Figure 9.2 Coefficient of variation for good and bad SMEs.

is 0.1498, and for the bad SMEs is 0.0082. This means that creditworthiness is more concentrated among good SME's.

9.4   Model building

Statistical credit scoring models try to predict the probability that a loan applicant or existing borrower will default over a given time horizon, usually of one year. According to the Basel Committee on Banking Supervision (BCBS) banks

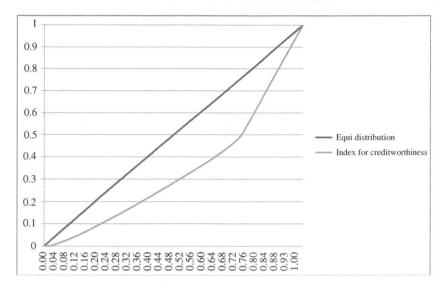

Figure 9.3 Creditworthiness index distribution.

are required to measure the one- year default probability for the calculation of the equity exposure of loans (see BCBS, 2005). To construct a scoring model, historical data on the performance of all loans in the loan portfolio have to be statistically analysed to determine the borrower characteristics that are useful in predicting whether a loan will perform well or poorly. Hence, a well-designed model should result in a higher percentage of high scores for borrowers whose loans will perform well and a higher percentage of low scores for borrowers whose loans will perform poorly. In other words the model should be well calibrated. A well-calibrated model yields – in the ideal case – as many realized defaults as predicted by the model. Historically, discriminant analysis and logistic regression have been the most widely used methods for constructing scoring systems (see 1997a, 1997b; Hand *et al.*, 2000, Hand and Henley). In particular, Altman (1968) was the first to use a statistical model to predict default probabilities of companies calculating the z-score using a standard discriminant model. This model was for many years one of the most prominent models for the calculation of a borrower's credit risk and the first that aimed to objectify the credit risk evaluation of banks' borrowers. Besides this basic method, more accurate ones such as logistic regression, neural networks, smoothing nonparametric methods and expert systems have been developed and are now widely used for practical and theoretical purposes in the field of credit risk measurement (Hand and Hanley, 1997a, 1997b).

Having performed the exploratory analysis, we move on to a multivariate analysis by specifying a statistical model. We are trying to combine all the signals from the different explanatory variables to obtain an overall signal that indicates the reliability of each SME. In order to choose a model, we have to clarify the

Table 9.1   Maximum likelihood estimates of the parameters.

Variable	Estimate	z value	$Pr(>|z|)$
liabilities_ratio_pc	7.099	3.137	<0.0001
result_ratio_pc	−34.52	−5.099	<0.0001
duration time	1.790	4.732	<0.0001
number of employees	−0.022	−3.425	<0.0001

nature of the problem. It is clear that we have a predictive classification problem, as the response variable is binary and our aim is to predict whether an SME will be reliable or not. We will concentrate on logistic regression and classification trees, the methods most often used for predictive classification in general, and credit scoring in particular.

We choose to implement a logistic regression model using a forward selection procedure with a significance level of 0.05. To check the model, we try a stepwise procedure and a backward procedure and then verify that the obtained models are similar.

Table 9.1 describes the models obtained with both procedures. We have used the score chi-squared statistic in the forward procedure and the Wald chi-squared statistic in the backward procedure.

To check the overall quality of the final model, we calculate the likelihood ratio test for the final model against the null model. As the corresponding p-value of the test is lower than 0.0001, the null hypothesis is rejected, implying that at least one of the model coefficients in Table 9.1 is significant.

For only three explanatory variables besides duration time we obtain a p-value lower than 0.05. This means that the three explanatory variables selected using the stepwise procedure are significantly associated with the response variable and are useful in explaining whether a SME is reliable (solvency $= 0$) or not (solvency $= 1$).

Turning to classification tree models, we begin with the CHAID algorithm and the chi-squared impurity measure. To obtain a parsimonious tree, we use a level of 0.05 in the stopping rule. The total number of splitting variables in the final tree is 4: result_ratio_pc, trade_payable_ratio_pc, supplier_target_days and capital_tied_up_pc.

We now look at a tree model using the CART algorithm and the Gini impurity. For pruning, we calculate the misclassification rate on the whole data set using the penalty parameter $\alpha = 1$. This can be considered as the default choice, in the absence of other considerations. The results for CART, Chaid and Gini impurity are the same. The final tree is reported in Figure 9.4. We observe that:

- For the 168 SMEs with result_ratio_pc < -0.00106379, the average probability of default is 0.35710.

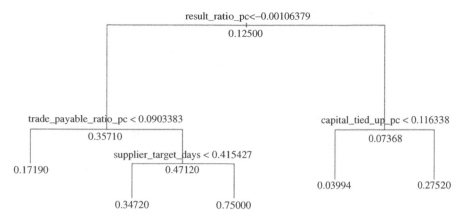

Figure 9.4 Final classification tree.

- For the 32 SMEs with result_ratio_pc < −0.00106379, trade_payable_ratio_pc > 0.0903383 and supplier_target_days > 0.415427, the estimated probability of default is 0.75.
- The segment characterised by capital_tied_up_pc < 0.116338 consists of 651 very good SMEs with a probability of default equal to 0.03994.

9.5 Model comparison

To help us choose a final model, we extend our performance analysis to include criteria based on loss functions. For all our models we begin by splitting the available data into a training data set, containing 75% of the observations, and a validation data set, containing the remaining 25%. We do this in a stratified way to maintain the proportions of good and bad SMEs.

After fitting each model on the training data set, we use it to classify the observations in the validation data set. This is done by producing a score and then using a threshold cut-off to classify those above the threshold as solvency = 1 and those below the threshold as solvency = 0. Finally, each model is evaluated by assessing the misclassification rate. In terms of cut-off we choose $p = 0.5$ as a 'majority rule' cut-off. With this cut-off, it turns out that the classification tree has a sensitivity of 0.20, a specificity of 0.99 and a proportion of correct classifications equal to 0.89. This compares to 0.09, 0.99 and 0.87 for the logistic regression. The two models are thus quite close in terms of performance.

Finally, to compare models with a threshold-independent measure of predictive performance the models, we computed the area under the concentration (AUC) curve and its confidence limits, by using h bootstrapped confidence intervals. Table 9.2 shows that the best model in AUC terms is the logistic regression. Having said that, the differences are rather slight.

Table 9.2   Model comparison.

Predictive model	AUC (95% confidence interval, bootstrap percentile method)
Classical logistic regression	0.8505 (0.81, 0.88)
Classical classification tree	0.8039 (0.75, 0.84)

9.6   Summary report

1. **Context.** This case study is concerned with credit scoring for SMEs based on balance sheet data. It may also be applied to situations where the objective is to score the past behaviour of an individual or company in order to plan a future action on the same individual or company. The score can be used to evaluate credit reliability or, similarly, customer loyalty. Furthermore, it can be used to select clients in order to maximise the return on an investment.

2. **Objectives.** The aim of the analysis is to construct a scoring rule that attaches a numerical value to each SME.

3. **Organisation of the data.** The data is organized in terms of financial ratios. The target variable is binary.

4. **Exploratory data analysis.** This phase was conducted using correlation measures.

5. **Model specification.** The objective of the analysis suggested a predictive model, able to find a rule that splits debtors into homogeneous categories and then attach to each category a score expressed as a probability of reliability.

6. **Model comparison.** The models were compared using statistical or scoring-based criteria. The goodness-of-fit comparison showed that logistic regression performed best, followed by classification trees.

7. **Model interpretation.** On the basis of model comparison, it seems that logistic regression does the best job for this problem. But classification trees models are not so inferior on the data set considered. The choice should also depend on how the results will be used. If decision makers are looking for hierarchical 'what if' rules, which classify clients into risk class profiles, then classification trees are very good. On the other hand, if they desire analytic rules, which attach an impact weight to each explanatory variable (measured by a regression coefficient or an odds ratio), then logistic regression is better.

CHAPTER 10

Predicting e-learning student performance

10.1 Objectives of the analysis

In this chapter we consider an e-learning context in place at Opera Multimedia, a multimedia publisher specialising in the production of e-learning content for the University of Pavia. Opera Multimedia offers multimedia courses that meet the educational needs of companies, universities and private users, designed to overcome the spatial and temporal constraints typical of traditional education. In particular, the new multimedia course combines the effectiveness of British Institutes teaching methods and the immediateness of the multimedia content. It is divided into three levels, consistent with the Common European Framework: Level 1 corresponds to A1 level and is devoted to beginners; Level 2 corresponds to A2 level and applies to pre-intermediate students; and Level 3 covers the B1 level and is designed for intermediate students.

Each level has 14–15 teaching units. Two of them – one in the middle and one at the end – revise topics previously discussed. Units focus on the main linguistic skills: reading, listening, writing and speaking. Listening activities, often neglected in online products, are particularly emphasised. The study of practice in the language are supported by images, animations and interactive exercises.

The course is available online, on CD-ROM, and on CD-ROM with tracking. Both the online and the CD-ROM versions have tracking for the monitoring of the activities performed and the results obtained. The CD-ROM with tracking is a novelty in the Italian market and represents a technological innovation that guarantees the highest flexibility while doing the course. In fact, students can access contents both online and on CD-ROM by connecting to the platform via web to send their progress of data. The student can supplement the course lessons in class at British Institutes centres and an online learning environment with in-depth teaching materials and tutoring service.

Designing an educational product means effectively blending all the elements involved in an e-learning activity: content, technology, cognitive aspects, and specific professional skills. The main objective for Opera Multimedia is to supply a measure of the relative importance of the exercises, to estimate the acquired knowledge for each student and to personalize the e-learning platform.

10.2   Description of the data

The available data come from an e-learning platform of the University of Pavia where students can learn English. Each of the 15 levels of the English course consists of 11 units (10 lessons and the final examination): Assessment, Dialogue, Glossary, Introduction, Listening 1, Listening 2, Pronunciation, Reading, Use of English, Video and Vocabulary. The course is divided into different types of exercises; some with evaluation (pronunciation, listening and assessment), others without (grammar). For each evaluation a score between 0 and 100 is given; the pass mark is 50.

For each student it is possible to collect that records a variable called 'status' the results achieved in the exercises. The possible values of the status variable are C (completed), I (uncompleted), F (failed), P (passed). We have eliminated from the initial table only the anomalous observations and the status I values. These account for 37 203 out of the 147 432 initial observations.

The data is structured into five tables: the demographic data related to the students enrolled in the course, the date and the initial and final time for every session in which a student is involved, the structure of the e-learning website and a transactional dataset of the lessons, as well as results from the final examination of each level. We have considered only the data related to the first course level that contained 463 students.

10.3   Exploratory data analysis

The first objective of the analysis is to supply a measure of the relative importance for each exercise, and the second is to predict the performance for each student. In this way we would like to personalize and to improve, following the previous results, the e-learning platform for a specific e-learning English course. To achieve these aims, we first consider the unsupervised statistical methods based on kernel density estimation (see Section 4.10 and, for more details, Figini and Giudici, 2008) for students' ecaluations.

The crucial point in the application of kernel density estimation is the choice of a bandwidth. This is a compromise between smoothing sufficiently to remove insignificant bumps and not smoothing too much to blur out real peaks. On the basis of our real data, we have compared two different methods to estimate the optimal smoothing parameter: that of Sheather and Jones (1991), and cross-validation. Figure 10.1 shows a histogram and the relative density estimation for exercise X10702 based on cross-validation. In particular, Figure 10.2 shows the kernel density estimator obtained using the smoothing parameter derived from the Sheather–Jones method. In our application the latter produces a fitted output which is the closest to the data.

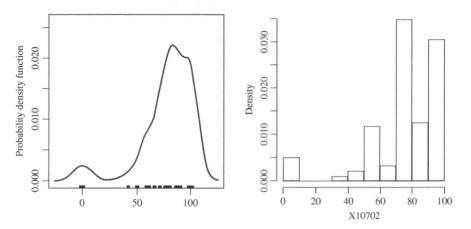

Figure 10.1 Density estimation for exercise X10702 evaluations with the cross-validation method.

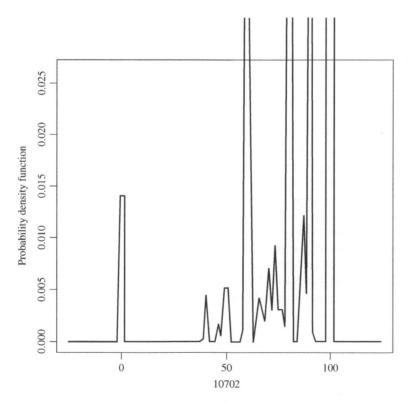

Figure 10.2 Density estimation for exercise X10702 evaluations with the Sheather – Jones method.

Table 10.1 Nonparametric pairwise comparison.

Exercise 1	Exercise 2	p-value
X10304	X10307	0.04
X10304	X10402	0.01
X10304	X10406	0.02
X10305	X10406	0.16
X10305	X10504	0.01
X10307	X10402	0.14
X10307	X10406	0.14
X10308	X10504	0.55
X10309	X10403	0.01
X10309	X10602	0.67
X10402	X10403	0.01
X10402	X10602	0.02
X10502	X10503	0.12
X10502	X10702	0.26

We have then supposed that it is possible to measure the relative importance for each exercise by means of evaluation density pairwise comparisons. We have chosen a nonparametric way to assess this: for each pair of exercises we have calculated the difference between the corresponding evaluation scores densities. Assuming as a null hypothesis that the two density functions for exercise f and exercise g are identical, it is possible to derive p-values with a bootstrap procedure that keeps h constant. The results are in Table 1. From Table 1 we have obtained that significantly different exercises are: X10304 and X10307, X10304 and X10402, X10304 and X10406, X10305 and X10504, X10309 and X10403, X10402 and X10403, X10402 and X10602.

Figure 10.3 shows, for exercises X10308 and X10504, a graphical comparison based on the confidence intervals obtained from the bootstrap procedure. The confidence interval is very close to the density estimate. This further suggests similarity between the two exercises considered. Therefore, in order to reduce the dimensionality of the predictive model described in the next section, either X10308 or X10504 could be dropped. On the other hand, Figure 10.4 shows for exercises X10309 and X10403 a statistically significant difference. This means that both exercises are important and should be kept in the model.

10.4 Model specification

To attain our second objective, of predicting the acquired knowledge for each student, we compare classical logistic regression models with the non-parametric additive models (see Hastie *et al.*, 2001). Our target binary variable is the status describing whether the student passes (status = 0) or fails (status = 1) the final exam. We consider 10 exercise evaluations as explanatory variables, selected

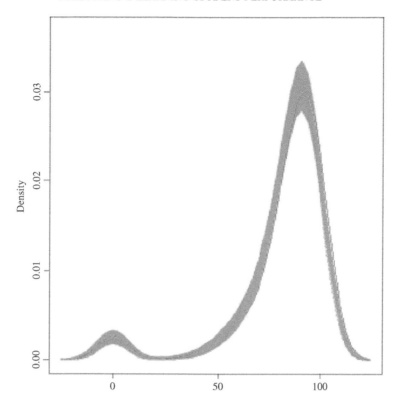

Figure 10.3 Graphical comparison between exercises X10308 and X10504.

according to the non-parametric pairwise comparison just described in Section 10.3. Table 10.2 shows the parameter estimates from logistic regression; note that only three exercises are significant for performance and for the final examination: X10308 (Pronunciation), X10309 (Listening) and X10702 (Comprehension).

We now compare the results in Table 10.2 with a non-parametric technique based on generalised additive models. One of the main reasons for using these is that they do not involve strong assumptions about the relationship that are implicit in standard parametric regression. The benefits in our application of an additive approximation are at least twofold. First, since each of the individual additive terms is estimated using a univariate smoother, the curse of dimensionality is avoided. Second, estimates of the individual terms explain how the dependent variable changes non-linearly with the corresponding explanatory variables. In fact generalised additive models extend the range of application of generaliaed linear models by allowing non-parametric smoothing predictors. In our application, Table 10.3 shows the generalised additive model outcome. For the estimation process, an iterative approach is used with backfitting algorithm. The significant exercises are X10308 (Pronunciation), X10309 (Listening), X10601 and X10602 (Comprehension). Comparing Table 10.2 and Table 10.3 note that they have X10308 and X10309 in common.

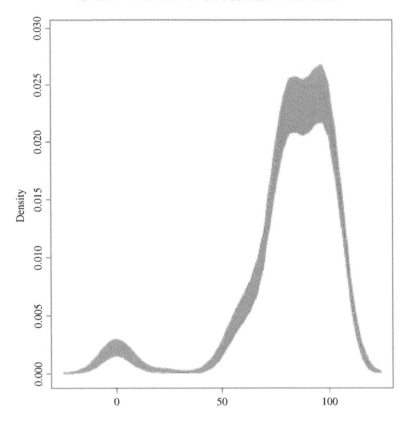

Figure 10.4   Graphical comparison between exercises X10309 and X10403.

Table 10.2   Estimation for the logistic regression model.

Variable	GLM logit	p-value
Intercept	−2.3121	0.0001
X10308	0.0396	0.0001
X10309	0.0291	0.0001
X10702	0.0344	0.0001

Table 10.3   Estimation for generalised additive model.

Spline	Chi-squared	Degrees of freedom
X10308	30.1602	3
X10309	7.8260	3
X10601	8.4466	3
X10602	10.3671	3

10.5   Model comparison

In order to choose the best predictive model between those that produce Tables 10.2 and 10.3, we report the confusion matrix. This is used as an indication of the properties of a classification (discriminant) rule. It contains the number of elements that have been correctly or incorrectly classified for each class. Its main diagonal shows the number of observations that have been correctly classified for each class, while the off-diagonal elements give the number of observations that have been incorrectly classified. If it is (explicitly or implicitly) assumed that each incorrect classification has the same cost in term of the acquired knowledge, one can calculate the total number of misclassifications as a performance measure. Table 10.4 shows the theoretical confusion matrix for a two-class classifier, as in our case. Given the context of our study, the entries in the confusion matrix have the following meaning: a is the number of correct predictions that a student will fail, b is the number of incorrect predictions that a student will fail, c is the number of incorrect predictions that a student will pass, and d is the number of correct predictions that a student will pass. Table 10.5 and 10.6 show the confusion matrices for the two models.

To obtain the figures in the tables we have used a cross-validation approach. We construct each model on a training sample and compare the models on a validation sample. The training sample (70%) and the validation sample (30%) are randomly selected. Comparing the two confusion matrices, we observe that

Table 10.4   Theoretical confusion matrix.

Observed/Predicted	Event	Non-event
Event	a	b
Non-event	c	d

Table 10.5   Confusion matrix for the logistic regression model.

	$P(Y = 0)$	$P(Y = 1)$
$O(Y = 0)$	59	22
$O(Y = 1)$	11	290

Table 10.6   Confusion matrix for the generalised additive model.

	$P(Y = 0)$	$P(Y = 1)$
$O(Y = 0)$	67	14
$O(Y = 1)$	6	285

the non-parametric model is better than logistic regression, as it leads to fewer misclassification errors. Based on the misclassification errors, we think that the non-parametric model selects with more accuracy specific exercises highly related to performance in the final examination. This empirical evidence leads to devote particular attention to specific exercises. This information may help our data provider to personalise the learning platform and to plan specific tutoring actions.

10.6   Summary report

1. **Context.** In this case study we analyse a real e-learning data set from the University of Pavia.
2. **Objectives.** We wish to provide a measure of the relative importance of exercises and to estimate the knowledge acquired by each student.
3. **Organization of the data.** Data is extracted from a large database containing a set of log files derived from the e-learning platform.
4. **Exploratory data analysis.** Exploratory data analysis was performed using a set of descriptive measures based on non-parametric techniques.
5. **Model specification.** We compare non-parametric additive models and parametric predictive models based on generalised linear models.
6. **Model comparison.** The methodology employed is based on a comparison between non-parametric statistical methods for kernel density classification and parametric models such as generalised linear models and generalized additive models.
7. **Model interpretation.** In this case study we have presented a novel approach to the analysis of e-learning platforms through the examination of student performance. Our proposal can be extended to other application areas such as credit risk, churn risk and, in general, risk measurement environments.

CHAPTER 11

Predicting customer lifetime value

11.1 Objectives of the analysis

In this chapter we consider statistical methods for lifetime value (LTV) estimation. Customer LTV measures the profit-generating potential, or value, of
a customer within the customer relationship management process. A customer
LTV model needs to be explained and understood before it can be adopted to
facilitate customer relationship management. LTV is usually taken to consist of
two independent components: tenure and value. Though modelling the value (or
equivalently, profit) component of LTV (which takes into account revenue, fixed
and variable costs) is a challenge in itself, our experience has shown that finance
departments, to a large extent, manage it reasonably well.

For a given customer there are three factors that need to be determined in order
to calculate LTV: the customer's value $v(t)$ over time $t > 0$; a model describing
the customer's churn probability over time; and a discount factor $D(t)$ which
describes how much each euro gained at some future time t is worth now. We
can then define $f(t) = -\,\mathrm{d}S(t)/\mathrm{d}t$ as the customer's instantaneous probability of
churn at time t, where $S(t)$ is the survival function. The quantity most commonly
modelled, however, is the hazard function $h(t) = f(t)/S(t)$. While $S(t)$ or $h(t)$
need to be estimated, $v(t)$ and $D(t)$ are usually known. We can write the explicit
formula for a customer's LTV, total value to be gained while the customer remains
active, as

$$\mathrm{LTV} = \int_{0}^{\infty} S(t)v(t)D(t)\,\mathrm{d}t;$$

The essence of a good LTV model is the estimation of $S(t)$ in a reasonable way.

Applied Data Mining for Business and Industry, 2e   P. Giudici, S. Figini
© 2009 John Wiley & Sons, Ltd

11.2   Description of the data

The subject of our case study is a well-known pay TV company. Considerations of confidentiality prevent us from giving accurate statements and information about this company; we shall instead make general statements and use normalized figures, and the company will simply be referred to as 'the company'. The main objective of such a company is to retain its customers in an increasingly competitive market, so it needs to know its customers' LTV and to carefully design appropriate marketing actions. Customers' contracts are renewed every year; if the customer does not withdraw, renewal takes place automatically. Otherwise the client churns. There are three types of churn events: people who withdraw from their contract in due time (i.e. less than 60 days before the due date); people who withdraw from their contracts outside due time (i.e. more than 60 days before or after the due date); and people who withdraw without giving notice, as is the case of bad payers. Churn events are classified into two different churn states: an 'exit' state for the first two classes of customers; and a 'suspension' state for the third.

Similarly to default (see Chapter 9), the causes of churn can be seen as consisting of a number of components: a static component determined by customer characteristics and the type of contract; a dynamic component that includes trends and the customers' contacts with the company call centre; a seasonal element, tied to contract duration; and external factors, including the performance of competitors and of the market in general. Currently the company uses a classification tree model that gives, for each customer, a churn probability (score). It is important for the company to be able to identify customers who are likely to leave and join a competitor. In business terms, predictive accuracy means being able to identify correctly those individuals who really will churn. Evaluation can be done using a confusion or cross-validation matrix. Static models, such as classification trees, show excessive influence of the contract deadline. It is therefore desirable to employ new methods to obtain a predictive tool which incorporates the fact that churn data are dynamic, that is, ordered in time.

The data available for our analysis includes information that can affect the distribution of the event time, such as demographic variables, variables related to the contract, payment, the contacts and geographical area of residence. The response variable, used as a dependent variable to construct predictive models, includes two different types of customers: those who during the survey are active and those who regularly cancel their subscription (EXIT status). We note that the target variable is observed 3 months after the extraction of the data set used for the model implementation phase, in order to verify correctly the effectiveness and predictive power of the models themselves. The data set contains 606 variables and a sample of 3500 observations (customers) and is composed of: socio-demographic information about the customers; information about their contractual situation and about its changes in time; information about contacting the customers (through the call centre, promotion campaigns, etc) and, finally, geo-marketing information (divided into census, municipalities and larger geographical sections information).

The variables regarding customers contain demographic information (age, gender, marital status, location, number of children, job and degree) and other information about customer descriptive characteristics: hobbies, PC possession at home, changes of address. The variables regarding the contract contain information about its chronology (signing date and starting date, time left before expiration date), its value (fees and options) at the beginning and at the end of the survey period, about equipments needed to use services (if they are rented, leased or purchased by the customer) and binary variables which indicate if the customer has already had an active, cancelled or suspended contract. There is also information about invoicing (invoice amount compared to different period of time: 2, 4, 8, 12 months). The variables regarding payment conditions include information about the type of payment of the monthly subscription (postal bulletin, account charge, credit card), as well as other information about the changes of the type of payment. The data set used for the analysis also includes variables which give information about the type of the services bought, about the purchased options, and about specific ad-hoc purchases, such as number and total amount of specific purchases during the last month and the last 2 months.

The variables regarding contacts with the customer contain information about any type of contact between the customer and the company (mostly through calls to the call centre). They include many types of calling categories (and relatives subcategories). They also include information about the number of questions made by every customer and temporal information, such as the number of calls made during the last month, the last two months and so on.

Finally, geo-marketing variables are present at large, and a great amount of work has involved their pre-processing and definition. Regardless of their provenience, all variables have gone through a pre-processing feature selection step aimed at reducing their very large number (equal to 606).

11.3 Exploratory data analysis

We first consider a non-parametric model, based on the Kaplan–Meier estimator, as described in Section 4.15. In order to construct a survival analysis model, we have created two variables: status (which distinguishes between active and non-active customers) and duration (an indicator of customer seniority). The first step in survival analysis is to plot the survival and hazard functions.

Figure 11.1 shows the estimated survival function and the related confidence interval for our data. Observe the variations in slope at different times. When the curve decreases rapidly we have high churn rates; when the curve decreases softly we have periods of 'loyalty'. The final jump is due to a distortion, caused by few data, in the tail of the lifecycle distribution. Figure 11.2 depicts the hazard function, which shows how the instantaneous risk rate varies over time. We observe two peaks, corresponding to times of greatest risk. Note that the risk rate is otherwise kept almost constant throughout the life cycle. Of course there is a peak at the end, corresponding to the phenomenon observed in Figure 11.1.

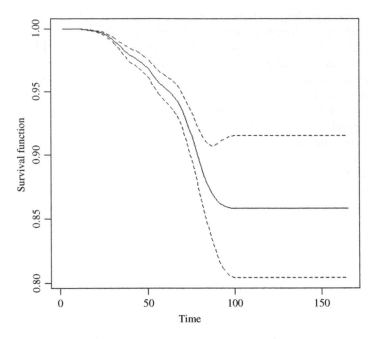

Figure 11.1   Survival function estimation based on the Kaplan–Meier estimator.

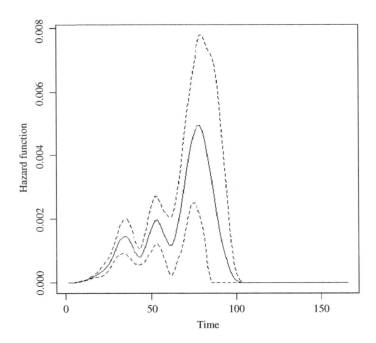

Figure 11.2   Estimated hazard function.

A very useful piece of information, in business terms, is customer life expectancy. This can be obtained as a sum over all observed event times,

$$\sum_{j=1}^{T} S(t_j) \times (t_j - t_{j-1}),$$

where $S(t_j)$ is the estimated survival function at the jth event time, obtained using the Kaplan–Meier method, and t is a duration indicator.

11.4 Model specification

In order to estimate the churn risk we use the Cox model described in Section 4.15. For the data at hand, the number of variables available after pre-processing is 24. These variables can be grouped into three main categories, according to the sign of their association with the churn rate, represented by the hazard ratio:

- variables that show a positive association (e.g. wealth of the geographic regions, quality of the call centre service, the sales channel);
- variables that show a negative association (e.g. number of technical problems, cost of service bought, payment method);
- variables that have no association (e.g. equipment rental cost, age of customer, number of family components).

To better interpret these associations we considered the value of the hazard ratio for different covariate values. For example, for the variable indicating number of technical problems we compared the hazard function for those that have called at least once with those that have not made any calls. As the resulting ratio turns out to be equal to 0.849, the risk of churning is lower for callers than for non-callers. The output of the Cox model is a new survival probability. In particular, this takes into account the multivariate relationships among the covariates, the target variable and the duration time. More precisely, the survival function is computed as

$$S(t, X) = S_0(t) \times \exp\left(\sum_{i=1}^{p} \beta_i X_i\right).$$

Figure 11.3 compares the survival curve obtained without covariates with the curve adjusted for the presence of covariates. Covariates have a large effect on survival times: up to 2 years of lifetime (24 months), the Cox survival curve (plotted with '+' symbols) is greater with respect to the baseline (solid curve). After 2 years the Cox survival probability declines abruptly and is much lower for the remaining lifespan. Once a Cox model has been fitted, it is advisable to produce diagnostic statistics, based on the analysis of residuals, to verify if the hypotheses underlying the model are correct.

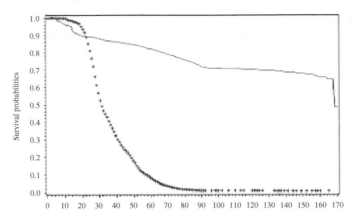

Figure 11.3 Comparison between survival functions, before and after covariate adjustment.

11.5   Model comparison

In order to create models to estimate the LTV for each customer, we employ the results from survival models. Survival analysis is useful for quantifying, in precise monetary terms, how much is gained or lost by replacing different customers with others characterized by different survival probabilities. For instance, how much is gained/lost if 0.08 of the clients, say, switch from buying service A to buying service B. Or, similarly, the relative gains when a certain percentage of clients changes method of payment (e.g. moving between payment by invoice, credit card and direct debit).

A simple way to quantify gains and losses is to calculate the area between two survival curves, as shown in Figure 11.4. Suppose the two survival curves correspond to two different services bought, which we will call black and grey, corresponding to the colours of the two curves. In order to determine exactly the area in Figure 11.4 we need to specify an amount of time ahead, say, 13 months. In Figure 11.4, the difference between survival probabilities 13 months after the customers became active is equal to 0.078. This value should be multiplied by the difference in business margin between the two methods of payment, as given, for example, by the difference in costs. Such costs can be described by a gain/loss table as in Table 11.1. A value of A is the relative gain if the client switches from PO to CC; similarly, B and C correspond respectively to relative gains from switching from PO to BA, and from CC to BA where PO = postal order, CC = payment through credit card, BA = payment via a bank account.

If we assume that we start with an acquired client base of 1000 customers in both categories (product black buyers and product grey buyers), the results say that, after 13 months we will remain with 934 black and 856 grey. If the finance department tells us that product black is worth €10 and product grey €20, then, after 13 months, we lose €660 for black churners and €2880 for grey churners. In other words, the priority of the marketing department should be to construct

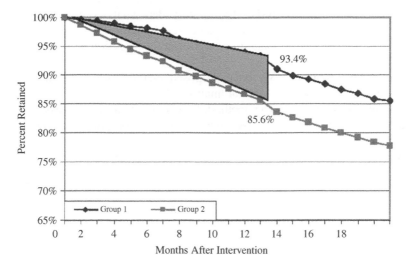

Figure 11.4 Evaluation of gains/losses by comparing survival curves.

Table 11.1 Evaluation of gain/losses by comparing survival curves.

	PO	CC	BA
PO		A	B
CC			C
BA			

targeted campaigns for grey product clients. From a different perspective, if black and grey correspond to two different sales channels for the same product, or to two different geographical areas, it is clear that the black channel (or area) is much better in terms of customer retention. Often promotional campaigns are conducted by looking only at increasing the customer base. Our results show that the number of captured clients should be traded off against their survival or, better, LTV profile.

11.6   Summary report

1. **Context.** In this case study we have analysed a real data set to predict customer lifetime value.
2. **Objectives.** We wish to estimate a measure of customer lifetime value for each customer on the basis of the duration time and a set of covariates.
3. **Organisation of the data.** Data were extracted from a large database coming from a pay-tv company.

4. **Exploratory data analysis.** Exploratory data analysis was performed by means of univariate and bivariate analyses. Specifically, in this chapter we applied the exploratory data process for outlier detection and variable selection.

5. **Model specification.** We have proposed methods to modelling customer lifetime value based on survival analysis.

6. **Model comparison.** The methodology employed was assessed using a cross-validation approach. We considered a set of predictive measures to choose the best model.

7. **Model interpretation.** The proposed models suggest ways to estimate a profitability score for each customer.

CHAPTER 12

Operational risk management

12.1 Context and objectives of the analysis

Recent legislation and market practices make it important to measure the risks arising as a result of management decisions. For instance, the Basel II capital accord, published by Basel Committee on Banking Supervision (BCBS, 2001), requires financial institutions to measure operational risk, defined as 'the risk of loss resulting from inadequate or failed internal processes, people and systems or from external events'. Another important standard is the recently published ISO 17799 which establishes the need for risk controls aimed at preserving the security of information systems. Finally, Publicly Available Specification PAS 56, in setting criteria that need to be met in order to maintain the business continuity of IT-intensive companies, also calls for the development of statistical indicators for monitoring the quality of business controls in place. In this chapter we focus on the Basel accord, while keeping in mind that what is developed here in a banking context can be extended to the general enterprise risk management framework (see Bonafede and Giudici, 2007).

The Bank of International Settlements (BIS) is the world's oldest financial institution; its main purpose is to encourage and facilitate cooperation among central banks (BCBS, 2001). In particular, the BIS established a commission, the Basel Committee on Banking Supervision, to formulate broad supervisory standards, guidelines and formulate best practice recommendations. The ultimate purpose of the BCBS is to prescribe capital adequacy standards for all internationally active banks.

In 1988 the BCBS issued one of the most significant international regulations impacting on the financial decision making of banks, the Basel accord. This was later replaced by the New Accord on Capital Adequacy, or Basel II (BCBS, 2001). This new framework, developed to ensure the stability and soundness of financial systems, was based on three 'pillars': minimum capital requirements, supervisory review and market discipline. What was important about the new accord was that it identified operational risk as a distinct category of risk. In fact, it was only with the new accord that the Risk Management Group of the Basel Committee proposed the current definition of operational risk, stated in the

Applied Data Mining for Business and Industry, 2e   P. Giudici, S. Figini
© 2009 John Wiley & Sons, Ltd

opening paragraph of this section. The Risk Management Group also provided a standardised classification of operational losses into eight business lines and seven event types, giving 56 possible categories.

The aims of operational risk measurement (Alexander, 2003; King, 2001; Cruz, 2002) are twofold. On one hand, there is a prudential aspect which involves setting aside an amount of capital to cover unexpected losses. This is typically done by estimating a loss distribution and deriving functions of interest from it, such as the value at risk (VaR). On the other hand, there is a managerial aspect, for which the issue is to rank operational risks in an appropriate way, say from high priority to low priority, so to individuate appropriate management actions directed at improving preventive control of such risks. In general, the measurement of operational risks leads to the measurement of the efficacy of controls in place at a specific organisation: the higher the operational risks, the worse such controls.

The complexity of operational risks and the newness of the problem have driven international institutions, such as the BCBS, to define conditions that sound statistical methodologies should satisfy to build and measure operational risk indicators.

12.2   Exploratory data analysis

Operational risk is composed of two elements: the frequency of loss events over a given period of time (frequency) and the mean monetary impact of the loss over the given period (severity). The following sources of information are generally used to determinate the operational value at risk:

- **Historical data.** In its simplest form this is a table whose rows contain information about the loss, such as the amount, the date of occurrence, the organisational unit identifying the loss, the amount recovered by the insurance and an indication of the business line and event type.
- **Expert opinion.** This is risk evaluations made by experts (branch heads, area heads, etc.) about the activity performed by the bank.
- **External databases.** These are typically consortium databases which aggregate the operational losses of all the banks taking part in the consortium. Here we shall refer to the Data Base Italiano delle Perdite Operative (Italian Database of Operational Losses, DIPO) consortium.

It is reasonable to expect some missing values among the events that constitute the universe of possible losses. Certain events have never happened in many situations or have never been registered either internally or at DIPO level. Sometimes experts do not manage to make evaluations. Thus some categories may have all three sources of information, while others have only two, one or even no sources.

The various sources of information are supplemented by a process of scaling. This makes it possible to compare and aggregate different databases, at least in terms of analysis. For example, we expect the DIPO, being an aggregation of the losses of several banks, to have loss amounts higher than the internal loss amounts. On the other hand, experts' opinions are formulated in specific areas, so their values may be rather different and, possibly, be comparable with internal losses.

In the DIPO case, scaling is done by dividing the losses in every category by a constant which is equal to the ratio between the total DIPO losses and the total internal losses. The same method can be applied to the frequency, so that we obtain the severity as a ratio. Obviously, in order for this scaling to be effective, the bank should have the same features as DIPO; given that DIPO gathers information from a set of banks in the same country, this requirement should be satisfied.

Another data organisation activity typically performed at the beginning of operational risk modelling is mapping. This is often a preliminary to the survey of expert opinion. By mapping we mean the identification, coding and schematization of all the activities carried out by a bank. This way, the organisation can be seen as an aggregation of productive processes, which are articulated as subprocesses, phases and subphases. The structure can be more or less complex according to the number of hierarchical levels considered. At the most elementary level, connected risks are identified. We talk about a risk when, in a given activity, there is a non-zero probability of incurring a loss.

Once mapping has been done, whether in the manner described or in a simpler manner starting with an exhaustive list of risk events (regardless of the processes), the expert must describe how the risk is perceived in terms of frequency and severity. Such opinions must refer to the Basel accord business lines and events types, so that each perceived loss can be mapped to a risk category.

The analysis of expert opinion is usually the exploratory analysis that is run in operational risk measurement. For the data set we have analysed, originating from a medium-sized Italian bank, we have noticed that experts agree about the distribution of the hazardousness across the losses, but not about their absolute dimension. This may be because different business areas have various levels of operational risk, as perceived by the experts. A scale factor could solve this problem, which will thus distinguish the various evaluations. Some experts may consider a loss of €10 000 as more or less insignificant, while for others it is very important.

Once expert opinions have been gathered, they are usually analysed with simple frequency distributions and graphs, one for each category. Aggregating the risks at the category level, we will obtain a database with 56 frequency × severity entries, one for every business line and event type. The product of the frequency and the severity gives the perceived loss for each category, and the sum of all such losses gives a qualitative (perceived) estimate of the total loss of the bank for every category.

Missing data is a problem that often occurs with expert opinions. This is sometimes due to the lack of evaluations by a few experts, sometimes due to the lack of a complete mapping activity of the process, which prevents the identification of some risks. In our case only two area experts have been asked. The mapping activity of the process, and the consequent identification of the risks, is exclusively concerned with the payment and settlements business area. To bypass the first missing-data problem we used the dimension of a business area (number of area branches related to the total number) to assess the quality of the data at our disposal. Therefore, for instance, since the payment and settlements area represents about the 17% of the total losses, the estimate of the total value on the basis of expert opinions has been obtained by dividing this value by 0.17. The idea to weight the evaluations by the number of branches is equivalent to the hypothesis that, at the branch level, the risk is almost identical, so that the risk of a broader area depends on its number of branches.

Another way to weight and scale opinions is to use key risk indicators (KRIs), variables which have a certain association with the losses and act as 'indicators' of the risk. Such variables are identified during the process mapping and monitored throughout the year. We decided to correct our experts' opinions using the KRIs available. The size of the correction depends on the significance of the KRI. For this purpose every loss, pertaining to the ith risk of the jth category evaluated by the kth expert, is multiplied by a factor given by

$$F_{i,j,k} = \begin{cases} 1 + 0.3R_{i,j}, & \text{if } \overline{K}_{i,j,k} > 0.66, \\ 1, & \text{if } 0.33 < \overline{K}_{i,j,k} < 0.66, \\ 1 - 0.3R_{i,j}, & \text{if } \overline{K}_{i,j,k} > 0.33, \end{cases}$$

where $\overline{K}_{i,j,k}$ is the average KRI associated with the ith risk of the jth category calculated on branches belonging to the area of the kth expert; and $R_{i,j}$ is the ratio between the deviation of the means of the area and total deviation relative to the ith risk and to the jth category.

The R^2 index represents the share of deviance that the classification is able to explain in terms of areas. This way, the amount of the correction is directly proportional to the significance of the KRI. If a KRI is not very discriminating from area to area, R^2 is small and so is the correction size. Conversely, if the KRI is appreciably different from area to area, the F factor will be larger.

12.3   Model building

Statistical models for operational risk can be grouped into two main categories: 'top-down' and 'bottom-up' methods. In top-down methods, risk estimation is based on macro data without identifying the individual events or the causes of losses. Operational risks are measured and covered centrally, so local business units are not involved in the measurement and allocation process. Top-down methods include the Basic Indicator Approach (Yasuda, 2003; Pézier, 2002) and the Standardised Approach (Cornalba and Giudici, 2004; Pézier, 2002), where risk is computed as a certain percentage of the variation of some variable, such as gross income, considered as a proxy for company performance. This approach is suitable for small banks, which prefer a methodology that is cheap to implement and easy to implement.

Bottom-up techniques, on the other hand, use individual events to determine the source and amount of operational risk. Operational losses can be divided into levels corresponding to business lines and event types, and the risks are measured at each level and then aggregated. These techniques are particularly appropriate for large banks and those operating at the international level, since they can afford the implementation of sophisticated methods which sensitive to the bank's risk profile. Advanced Measurement Approaches (AMA) belong to this class (BCBS, 2001). Under the AMA, the regulatory capital requirement will equal the risk measure generated by the bank's internal operational risk measurement system using the quantitative and qualitative criteria set by the BCBS. This is an advanced approach as it allows banks to use external and internal loss data as well as internal expertise (Giudici and Bilotta, 2004).

Statistical methods for operational risk management in the bottom-up context have recently been developed. One main approach has emerged: the actuarial approach. This method uses actual loss data to estimate the probability distribution of the losses. The most popular methods (King, 2001; Cruz, 2002; Frachot et al., 2001; Dalla Valle et al., 2008) are based on extreme value distributions. Another line of research suggests the use of Bayesian models (Yasuda, 2003, Cornalba and Giudici, 2004; Fanoni et al., 2005; Dalla Valle and Giudici, 2008). The main disadvantage of actuarial methods is their backward-looking perspective; their estimates are based entirely on past data. Furthermore, it is often the case, especially for smaller organisations, that for some business units there are no loss data at all. Regulators thus recommend developing models that can take into account different data streams, not only internal loss data (see BCBS, 2001). These streams may be: self assessment opinions, usually forward looking; external loss databases, usually gathered through consortia of companies; and data on key performance indicators.

In the actuarial model, loss events are assumed independent and, for each one, it is assumed that the total loss in a given period (e.g. one year) is obtained as the sum of a random number (N) of impacts (X_i). In other words, for the jth event the loss is

$$L_j = \sum_{i=1}^{N_j} X_{ij}.$$

Usually the distribution of each j-specific loss is obtained from the specification of the distribution of the frequency N and the mean loss or severity S. The convolution of the two distributions leads to the distribution of L (typically via a Monte Carlo estimation step), from which a function of interest, such as the 99.9 percentile (the value at risk) can be derived.

The scorecard approach is based on the so-called self assessment, which is based on the experience and the opinions of a number of internal 'experts' of the company, who usually correspond to a particular business unit. An internal procedure of control self assessment can periodically be done by means of questionnaires. Questionnaires can be submitted to risk managers (experts), and give information such as the quality of internal and external control systems of the organisation based on their experience over a given period. In a more sophisticated version, experts can also assess the frequency and mean severity of the losses for such operational risks (usually in a qualitative way). Self assessment opinions can be summarised and modelled so as to obtain a ranking of the different risks, and a priority list of interventions in terms of improvement of the related controls.

In order to derive a summary measure of operational risk, perceived losses contained in the self-assessment questionnaire can be represented graphically (e.g. in a histogram), leading to an empirical non-parametric distribution. Such a distribution can be employed to derive a function of interest, such as the VaR.

Scorecard models are useful for prioritising interventions in the control system, so as to effectively reduce the impact of risks, ex ante and not a posteriori, by allocating capital for example (corresponding to the VaR).

In Giudici (2008) a methodology aimed at summarising concisely and effectively the results of a self-assessment questionnaire has been proposed. In the next section we shall show how it can be applied.

12.4 Model comparison

Suppose that we are given 80 events at risk (this is the order of magnitude employed in a typical operational risk management analysis in the banking sector). The events can be traced to the four main causes of operational risk: People, processes, systems and external events. First of all a sample of banking professionals (from headquarters and the local branches) is obtained. The aims of the questionnaire project are described in a group presentation. The nature and structure of each risk question will have been devised in a focus group discussion with the top manager of the bank. The result of this preliminary analysis is that each of the professionals is asked his/her opinion on the frequency, severity and effectiveness of the controls in place for each risk event. The number of possible frequency classes is equal to four: daily, weekly, monthly, and yearly. The number of severity classes depends on the size of the bank's capital, and will typically be six or seven, ranging from 'an irrelevant loss' to 'a catastrophic loss'. Finally there are three possible classes for the controls: not effective, to be adjusted, and effective.

Once the interviews have been gathered together, the aim is to assign a 'rating' to each risk event, based on the distribution of the opinions on the frequency, controls and severity. Giudici (2008) suggests using the median class as a location measure for each distribution, and the normalised Gini index as an indicator of the 'consensus' on such location measure. This results in three rating measures for each event, denoted by the conventional risk letters: A for low risk, B for medium risk, C for higher risk and so on. While the median is used to assign a single-letter measure, the Gini index is used to double or triple the letter, depending on the value of the index. For example, if the median of the frequency distribution of a certain risk type (e.g. theft and robbery) is yearly, corresponding to the lowest risk category, a letter A is assigned. Then, if all the interviewees agree on that evaluation (e.g. the Gini index is equal to zero), A becomes AAA; if the Gini index corresponds to maximum heterogeneity, then A remains A. Intermediate cases will receive an AA rating. The same approach will serve for the severity as well as for the controls, leading to a complete scorecard that can be used for intervention purposes. For visualisation purposes, colours are associated with letters, using a traffic-light convention: green (grey); corresponds to A; yellow (light grey) to B; red (dark grey) to C and so on.

Figure 12.1 presents the results from such scorecard model for a collection of risk events caused by to people (internal fraud) and external events (external fraud and losses from material activities). It turns out that event 1.2.6 should be given top priority for intervention, as controls are not effective, and both frequency and severity are yellow. Other events at risk include 2.2.1 and 2.2.4, which have a high frequency and medium quality controls. We observe that opinion on the severity is usually considered second in priority determination as it typically concerns a mean value which cannot be modified by the action of controls.

While scorecard methods typically use self-assessment data, actuarial models use internal loss data. The disadvantage of these approaches is that they consider only one part of the statistical information available to estimate operational risks. Actuarial methods rely only on past loss data and therefore do not consider important information on the perspective and the evolution of the company; on the other hand, scorecard methods are based only on perceived data (are forward-looking) and therefore do not reflect past experience very well.

A further problem is that, especially for rare events, a third data stream may be considered: external loss data. This source of data is made up of pooled records of losses, typically higher than a certain value (e.g. €5000), collected by an appropriate association of banks.

It thus becomes necessary to develop a statistical methodology that is able to merge three different data streams in an appropriate way, while maintaining simplicity of interpretation and predictive power. Here we shall propose a flexible non-parametric approach that can achieve this objective (see Giudici, 2008). Such an approach can be justified within a non-parametric Bayesian context.

Our approach considers, for each event, data on all losses that have occurred in the past as well as the expected self assessment losses for the next period. The latter is counted as one data point, typically higher than actual losses, even when calculated as a mean loss rather than as a worst case loss. Putting together the self-assessment data point with the actual loss data points, we obtain an integrated loss distribution, from which a VaR can be calculated. Alternatively, to take the loss distributions more correctly into account, a Monte Carlo simulation can be based on the given losses, leading to a (typically higher) Monte Carlo VaR, parallel to what is usually done in the actuarial approach.

In Figure 12.2 we compare, for a real database, the VaR obtained using a pure self-assessment approach, with the actuarial VaRs (both historical and Monte Carlo based) and the integrated (Bayesian) VaR (both simple and Monte Carlo). For reasons of predictive accuracy, we construct all methods on a series of data

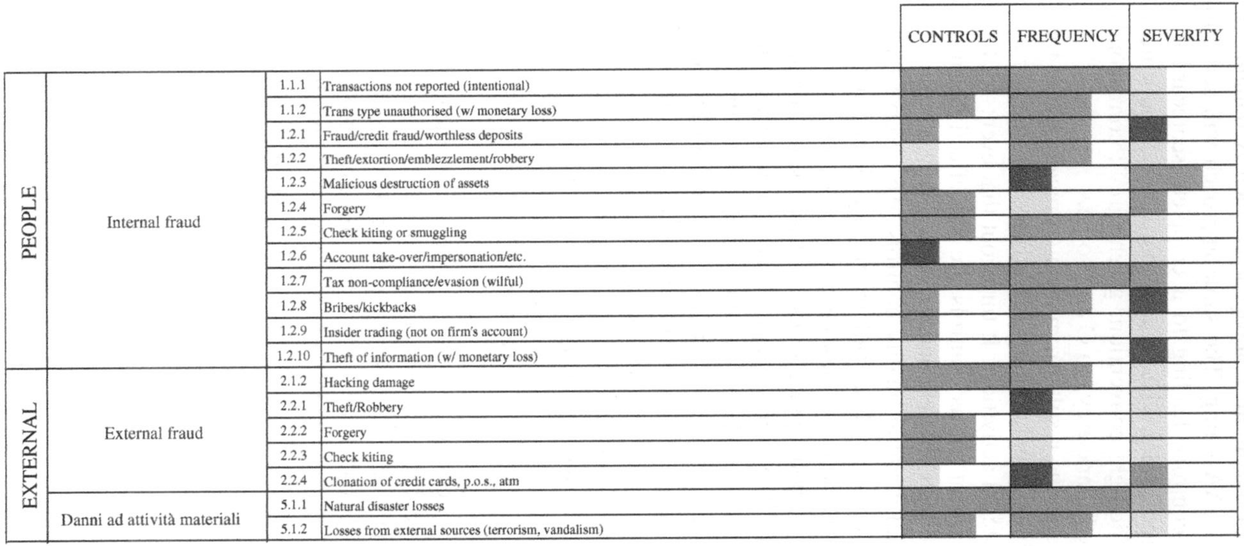

Figure 12.1 Example of results from our proposed scorecard model.

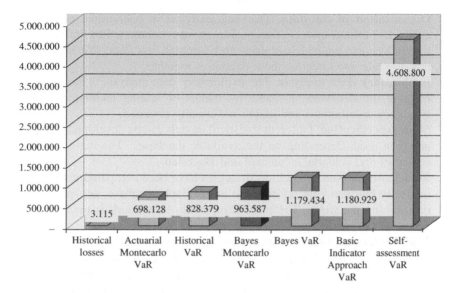

Figure 12.2 Example of results from our integrated scorecard model.

points updated at the end of 2005; calculate the VaR for 2006 (possibly integrating it with the self assessment available opinions for 2006) and compare the VaR with the actual losses for 2006. We also calculate the VaR that would be obtained under the simple Basic Indicator Approach (BIA) suggested by Basel II. The BIA amounts to calculating a flat percentage (15%) of a relevant indicator (such as the gross income), without further elaboration. It turns out that both our proposed models (Bayes VaR and Bayes Monte Carlo) lead to an allocation of capital (represented by the VaR) lower than the BIA and higher than the observed losses. Although these results are achieved by the actuarial models as well (historical and actuarial Monte Carlo), we believe that a prudential approach, as expressed by our proposal, is more sound, especially over a longer time horizon.

12.5 Summary conclusions

1. **Context.** We have shown how to develop efficient statistical methods for measuring the performance of business controls, through the development of appropriate operational risk indicators.
2. **Objectives.** The advantages of measuring operational risk appropriately are twofold. On the one hand, there is a prudential aspect which involves setting aside an amount of capital that can cover unexpected losses. On the other hand, there is a managerial aspect for which the issue is to rank operational risks in an appropriate way, say from high priority to low priority, so as to identify appropriate management actions directed at improving preventive controls on such risks.

3. **Organisation of the data.** The data analysed in operational risk usually comes from three sources: expert opinion, often corrected by means of key risk indicators; internal loss data; and external loss data. We have shown how to scale and integrate these databases together.

4. **Exploratory data analysis.** The exploratory data analysis is based on first merging the databases available by means of scaling techniques. Simple frequency distributions and graphs are usually obtained.

5. **Model specification.** Different statistical models can be developed, compared and then used, depending on the available databases. The main modelling strategies are: actuarial, scorecard and Bayesian.

6. **Model comparison.** Modelling strategies have been compared not only theoretically but also with reference to an available database, according to the data mining paradigm. The main comparison benchmark is predictive performance (known in this context as backtesting).

7. **Model interpretation.** The measurement of operational risk has been made compulsory by recent legislative changes. We have shown that, when models are transparent and interpretable, the measurement of operational risk leads to the measurement of the efficacy of controls in place at a given organisation: the higher the operational risks, the worse the controls.

References

Agrawal, R., Mannila, H., Srikant, R., Toivonen, H. and Verkamo, A.I. (1996) Fast discovery of association rules. In U.M. Fayyad, G. Piatetsky-Shapiro, P. Smyth, and R. Uthurusamy (eds) *Advances in Knowledge Discovery and Data Mining*. AAAI/ MIT Press, Menlo Park, CA.

Agresti, A. (1990) *Categorical Data Analysis*. John Wiley & Sons, Inc., New York.

Akaike, H. (1974) A new look at statistical model identification. *IEEE Transactions on Automatic Control* **19**, 716–723.

Alexander, C. (ed.) (2003) *Operational Risk: Regulation, Analysis and Management*. Financial Times Prentice Hall, London.

Allison, P. (1995) *Survival Analysis Using the SAS System: A Practical Guide*. SAS Institute, Cary, NC.

Altman, E. (1968) Financial ratios, discriminant analysis and the prediction of corporate bankruptcy. *Journal of Finance* **23**(4).

Azzalini, A. (1992) *Statistical Inference: An Introduction Based on the Likelihood Principle*. Springer-Verlag, Berlin.

Barnett, V. (1974) *Elements of Sampling Theory*. English Universities Press, London.

Basel Committee on Banking Supervision (2001) Working Paper on the Regulatory Treatment of Operational Risk. www.bis.org.

Basel Committee on Banking Supervision (2005) Amendment to the Capital Accord to Incorporate Market Risks. Basel.

Beaver, W. (1966) Financial ratios as predictors of failure, *Journal of Accounting Research*, **4**(supplement): 71–111.

Benzécri, J.-P. (1973) *L'Analyse des données*. Dunod, Paris.

Bernardo, J.M. and Smith, A.F.M. (1994) *Bayesian Theory*. John Wiley & Sons, Inc., New York.

Berry, M. and Linoff, G. (1997) *Data Mining Techniques for Marketing, Sales, and Customer Support*. John Wiley & Sons, Inc., New York.

Berry, M. and Linoff, G. (2000) *Mastering Data Mining*. John Wiley & Sons, Inc., New York.

Bickel, P.J and Doksum, K.A. (1977) *Mathematical Statistics*. Prentice Hall, Englewood Cliffs, NJ.

Bishop, C. (1995) *Neural Networks for Pattern Recognition*. Clarendon Press, Oxford.

Bollen, K.A. (1989) *Structural Equations with Latent Variables*. John Wiley & Sons, Inc., New York.

Bonafede, E.C. and Giudici, P. (2007) Bayesian networks for enterprise risk assessment. *Physica A* **382**: 22–28.

Breiman, L., Friedman, J.H., Olshen, R. and Stone, C. J. (1984) *Classification and Regression Trees*. Wadsworth, Belmont, CA.

Brooks, S.P., Giudici, P. and Roberts, G.O. (2003) Efficient construction of reversible jump MCMC proposal distributions (with discussion). *Journal of the Royal Statistical Society, Series B* **65**: 1–37.

Cadez, I., Heckerman, D., Meek, C., Smyth, P. and White, S. (2000) Visualization of navigation patterns on a web site using model based clustering. In *Proceedings of the Sixth ACM SIGKDD International Conference on Knowledge Discovery and Data Mining*, Boston, MA.

Cassel, C.M. (2000) Measuring customer satisfaction on a national level using a superpopulation approach. *Total Quality Management* **11**(7): 909–915.

Cassel, C.M. and Eklöf, J. (2001) Modelling customer satisfaction and loyalty on aggregate levels: Experience from the ECSI pilot study. *Total Quality Management* **12** (7–8): 834–841.

Cassel, C., Eklöf, J., Hallissey, A., Letsios, A. and Selivanova I. (2002) The EPSI rating initiative. *European Quality* **9**(2): 10–25.

Castelo, R. and Giudici, P. (2003) Improving Markov chain model search for data mining. *Machine Learning* **50**: 127–158.

Chatfield, C. (1996) *The Analysis of Time Series: An Introduction*. Chapman & Hall, London.

Chava, S. and Jarrow, R. (2004) Bankruptcy prediction with industry effects. *Review of Finance* **8**(4): 537–569.

Cheng, S. and Titterington, M. (1994) Neural networks: A review from a statistical perspective. *Statistical Science* **9**: 3–54.

Christensen, R. (1997) *Log-Linear Models and Logistic Regression*. Springer-Verlag, Berlin.

Cifarelli, D.M. and Muliere, P. (1989) *Statistica Bayesiana*. Iuculano, Pavia.

Coppi, R. (2002) A theoretical framework for data mining: the 'information paradigm'. *Computational Statistics & Data Analysis* **38**: 501–515.

Cornalba, C. and Giudici, P. (2004) Statistical models for operational risk management. *Physica A*, 338: 166–172.

Cortes, C. and Pregibon, D. (2001) Signature-based methods for data streams. *Journal of Knowledge Discovery and Data Mining* **5**: 167–182.

Cowell, R.G., Dawid, A.P, Lauritzen, S.L. and Spiegelhalter, D.J. (1999) *Probabilistic Networks and Expert Systems*. Springer-Verlag, New York.

Cox, D.R. (1972) Regression models and life tables. *Journal of the Royal Statistical Society, Series B* **34**: 187–220.

Cox, D.R. and Wermuth, N. (1996) *Multivariate Dependencies. Models, Analysis and Interpretation*. Chapman & Hall, London.

Cressie, N. (1991) *Statistics for Spatial Data*. John Wiley & Sons, Inc., New York.

Cruz, M. (2002) *Modelling, Measuring and Hedging Operational Risk*. John Wiley & Sons, Ltd, Chichester.

Dalla Valle, L., Fantazzini, D. and Giudici, P. (2008) Copulae and operational risks. *International Journal of Risk Assessment and Management* **9**(3): 238–257.

Dalla Valle, L. and Giudici, P. (2008) A Bayesian approach to estimate the Marginal loss distributions in Operational Risk management. *Computational Statistics and data analysis* **52**: 3107–3127.

Darroch, J.N., Lauritzen, S.L. and Speed, T.P. (1980) Markov fields and log-linear models for contingency tables. *Annals of Statistics* **8**: 522–539.

De Ville, B. (2001) *Microsoft Data Mining, Integrated Business Intelligence for e-Commerce and Knowledge Management*. Digital Press, New York.

Dempster, A. (1972) Covariance selection. *Biometrics* **28**: 157–175.

Diggle, P.J., Liang, K. and Zeger, S.L. (1994) *Analysis of Longitudinal Data*. Clarendon Press, Oxford.

Dobson, A.J. (2002) *An Introduction to Generalized Linear Model*, 2nd edition. Chapman & Hall/CRC, Boca Raton, FL.

Edwards, D. (2000) *Introduction to Graphical Modelling* (2nd edition). Springer-Verlag, New York.

Efron, B. (1979) Bootstrap methods: Another look at the jackknife. *Annals of Statistics* **7**: 1–26.

Engelmann, B., Hayden, E. and Tasche, D. (2003) Measuring the discriminative power of rating systems, banking and financial supervision. Deutsche Bundesbank Discussion Paper no. 01/2003.

Fanoni, F., Giudici, P. and Muratori, G.M. (2005) *Il Rischio Operativo: Misurazione, Monitoraggio, Mitigazione*. Il Sole 24 Ore.

Figini, S. and Giudici, P. (2008) Statistical models for e-learning data. *Statistical Methods and Applications*. To appear.

Frachot, A., Georges, P. and Roncalli, T. (2001) Loss distribution approach for operational risk. Working Paper, Groupe de Recherche Opérationnelle du Crédit Lyonnais.

Frydenberg, M. and Lauritzen, S.L. (1989) Decomposition of maximum likelihood in mixed interaction models. *Biometrika* **76**: 539–555.

Gibbons, D. and Chakraborti, S. (1992) *Nonparametric Statistical Inference*. Marcel Dekker, New York.

Gilks, W.R., Richardson, S, and Spiegelhalter, D.J. (eds) (1996) *Markov Chain Monte Carlo in Practice*. Chapman & Hall, London.

Giudici, P. (2001a) Bayesian data mining, with application to credit scoring and benchmarking. *Applied Stochastic Models in Business and Industry* **17**: 69–81.

Giudici, P. (2008) Statistical models for risk governance of public services. In *Proceedings of the first joint meeting of the societancofone de classification and the data analysis group of the Italian statistical society*, pp. 19–22.

Giudici, P. and Bilotta, A. (2004) Modelling operational losses: A Bayesian approach. *Quality and Reliability Engineering International* **20**: 407–417.

Giudici, P. and Carota, C. (1992) Symmetric interaction models to study innovation processes in the European software industry. In L. Fahrmeir, B. Francis, R. Gilchrist and G. Tutz (eds), *Advances in GLIM and Statistical Modelling*. Springer-Verlag, Berlin.

Giudici, P. and Castelo, R. (2001) Association models for web mining. *Journal of Knowledge Discovery and Data Mining* **5**: 183–196.

Giudici, P. and Green, P.J. (1999) Decomposable graphical Gaussian model determination. *Biometrika* 86: 785–801.

Giudici, P. and Passerone, G. (2002) Data mining of association structures to model consumer behaviour. *Computational Statistics and Data Analysis* **38**: 533–541.

Goodman L.A. and Kruskal W.H (1979) *Measures of Association for Cross Classification*. Springer-Verlag, New York.

Gower, J.C. and Hand, D. J. (1996) *Biplots*. Chapman & Hall, London.

Green, P.J., Hjort, N. And Richardson, S. (eds) (2003) *Highly Structured Stochastic Systems*. Oxford University Press, Oxford.

Greenacre, M. (1983). *Theory and Applications of Correspondence Analysis*. Academic Press, New York.

Greene, W.H. (2000) *Econometric Analysis*. Prentice Hall, Upper Saddle River, NJ.

Han, J. and Kamber, M (2001) *Data Mining: Concepts and Techniques*. Morgan Kaufmann, San Francisco.

Hand, D. (1997) *Construction and Assessment of Classification Rules*. John Wiley & Sons, Ltd, Chichester.

Hand, D.J. and Henley, W.E. (1997a) Some developments in statistical credit scoring. In G. Nakhaeizadeh and C.C. Taylor (eds), *Machine Learning and Statistics: The Interface*, pp. 221–237. John Wiley & Sons, Inc., New York.

Hand, D.J. and Henley, W.E. (1997b) Statistical classifications method in consumer scoring: a review. *Journal of the Royal Statistical Society, Series A* **160**: 523–541

Hand, D.J., Blunt, G., Kelly, M.G. and Adams, M.N. (2000) Data mining for fun and profit. *Statistical Science* **15**: 111–131.

Hand, D.J., Mannila, H. and Smyth, P (2001) *Principles of Data Mining*. MIT Press, Cambridge, MA.

Hastie, T., Tibshirani, R., Friedman, J. (2001) *The Elements of Statistical Learning: Data Mining, Inference and Prediction*. Springer-Verlag, New York.

Heckerman, D. (1997) Bayesian networks for data mining. *Journal of Data Mining and Knowledge Discovery* **1**: 79–119.

Hoel, P.G., Port, S.C., and Stone, C.J. (1972) *Introduction to Stochastic Processes*. Waweland Press, Prospect Heights, IL.

Hougaard, P. (1995) Frailty models for survival data. *Lifetime Data Analysis* **1**: 255–273.

Jensen, F. (1996) *An Introduction to Bayesian Networks*. Springer-Verlag, New York.

Kaplan, E.L. and Meier, P. (1958) Nonparametric estimation from incomplete observations. *Journal of the American Statistical Association*, **53**: 457–481.

Kass, G.V. (1980) An exploratory technique for investigating large quantities of categorical data. *Applied Statistics* **29**: 119–127.

King, J. (2001) *Operational Risk. Measurement and Modelling*. John Wiley & Sons, Ltd, Chichester.

Kolmogorov, A.N. (1933) Sulla determinazione empirica di una leggi di probabilita. *Giornale dell'Istituto Italiano degli Attuari* **4**.

Lauritzen, S.L. (1996) *Graphical Models*. Oxford University Press, Oxford.

Mardia, K.V., Kent, J.T. and Bibby, J.M. (1979) *Multivariate Analysis*. Academic Press, London.

McCullagh, P. and Nelder, J.A. (1989) *Generalised Linear Models*. Chapman & Hall, New York.

Mood, A.M., Graybill, F.A. and Boes, D.C. (1991) *Introduction to the Theory of Statistics*. McGraw-Hill, Tokyo.

Neal, R. (1996) *Bayesian Learning for Neural Networks*, Lecture Notes in Statistics 118. Springer-Verlag, New York.

Nelder, J.A. and Wedderburn, R.W.M. (1972) Generalized linear models. *Journal of the Royal Statistical Society, Series B* **54**: 3–40.

Pézier, J. (2002) A constructive review of the Basel proposals on operational risk. ISMA Technical report, University of Reading.

Quinlan, R. (1993) *C4.5: Programs for Machine Learning*. Morgan Kaufmann, San Mateo, CA.

Ripley, B.D. (1996) *Pattern Recognition and Neural Networks*. Cambridge University Press, Cambridge.

Särndal, C.E. and Lundström, S. (2005) *Estimation in Surveys with Nonresponse*. John Wiley & Sons, Ltd, Chichester.

Schwarz, G. (1978) Estimating the dimension of a model. *Annals of Statistics* **62**: 461–464.

Searle, S.R. (1982) *Matrix Algebra Useful for Statistics*. John Wiley & Sons, Inc., New York.

Sheather, S.J. and Jones, M.C. (1991) A reliable data based bandwidth selection method for kernel density estimation. *Journal of the Royal Statistical Society, Series B* **53**: 683–690.

Shumway, T. (2001) Forecasting bankruptcy more accurately: A simple hazard model. *Journal of Business* **74**: 101–124.

Singer, J.D. and Willett J.B. (2003) *Applied Longitudinal Data Analysis: Modeling Change and Event Occurrence*. Oxford University Press, New York.

Siskos Y., Grigoroudis E., Zopounidis C. and Saurais O. (1998) Measuring customer satisfaction using a collective preference disaggregation model. *Journal of Global Optimization* **12**, 175–195.

Sobehart, J.R. and Keenan, S.C. (2001) A practical review and test of default prediction models. *RMA Journal*, 54–59.

Stein, R.M. (2005) The relationship between default prediction and lending profits: Integrating ROC analysis and loan pricing. *Journal of Banking and Finance* **29**(5): 1213–1236.

Tukey, J.W. (1977) *Exploratory Data Analysis*. Addison Wesley, Reading, MA.

Vapnik, V. (1995) *The Nature of Statistical Learning Theory*. Springer-Verlag, New York.

Vapnik, V. (1998) *Statistical Learning Theory*. John Wiley & Sons, Inc., New York.

Venables, W.N. and Ripley, B.D. (2002) *Modern Applied Statistics with S*, 4th edition. Springer-Verlag, New York.

Whittaker, J. (1990) *Graphical Models in Applied Multivariate Statistics*. John Wiley & Sons, Ltd, Chichester.

Yasuda, Y. (2003) Application of Bayesian inference to operational risk management. Technical report, University of Tsukuba.

Zadeh, L.A. (1977) Fuzzy sets and their application to pattern classification and clustering. In J. Van Ryzin (ed.), *Classification and Clustering*. Academic Press, New York.

Zanasi, A. (ed.) (2003) *Text Mining and its Applications*. WIT Press, Southampton.

Zucchini, W. (2000) An introduction to model selection. *Journal of Mathematical Psychology* **44**: 41–61.

Index

Printed and bound by CPI Group (UK) Ltd, Croydon, CR0 4YY

27/10/2024

14580157-0002